O9-AHV-839

TED

More Praise for *The Second Machine Age*

"Brynjolfsson and McAfee are right: we are on the cusp of a dramatically different world brought on by technology. *The Second Machine Age* is the book for anyone who wants to thrive in it. I'll encourage all of our entrepreneurs to read it, and hope their competitors don't."
—Marc Andreessen, cofounder of Netscape and Andreessen Horowitz

"What globalization was to the economic debates of the late 20th century, technological change is to the early 21st century. Long after the financial crisis and great recession have receded, the issues raised in this important book will be central to our lives and our politics."
—Lawrence H. Summers, Charles W. Eliot University Professor at Harvard University

"Technology is overturning the world's economies, and *The Second Machine Age* is the best explanation of this revolution yet written."
—Kevin Kelly, senior maverick for *Wired* and author of *What Technology Wants*

"Brynjolfsson and McAfee take us on a whirlwind tour of innovators and innovations around the world. But this isn't just casual sightseeing. Along the way, they describe how these technological wonders came to be, why they are important, and where they are headed."
—Hal Varian, chief economist at Google

"In this optimistic book Brynjolfsson and McAfee clearly explain the bounty that awaits us from intelligent machines. But they argue that creating the bounty depends on finding ways to race *with* the machine rather than racing *against* the machine. That means people like me need to build machines that are easy to master and use. Ultimately, those who embrace the new technologies will be the ones who benefit most."
—Rodney Brooks, chairman and CTO of Rethink Robotics, Inc

"New technologies may bring about our economic salvation or they may threaten our very livelihoods . . . or they may do both. Brynjolfsson and McAfee have written an important book on the technology-driven opportunities and challenges we all face in the next decade. Anyone who wants to understand how amazing new technologies are transforming our economy should start here."

—Austan Goolsbee, professor of economics at the University of Chicago Booth School of Business and former chairman of the Council of Economic Advisers

"After reading this book, your world view will be flipped: you'll see that collective intelligence will come not only from networked brains but also from massively connected and intelligent machines. In the near future, the best job to have will be the one you would do for free."

—Nicholas Negroponte, cofounder of the MIT Media Lab, founder of One Laptop per Child, and author of *Being Digital*

"*The Second Machine Age* helps us all better understand the new age we are entering, an age in which by working with the machine we can unleash the full power of human ingenuity. This provocative book is both grounded and visionary, with highly approachable economic analyses that add depth to their vision. A must-read."

—John Seely Brown, coauthor of *The Power of Pull* and *A New Culture of Learning*

"Brynjolfsson and McAfee do an amazing job of explaining the progression of technology, giving us a glimpse of the future, and explaining the economics of these advances. And they provide sound policy prescriptions. Their book could also have been titled Exponential Economics 101—it is a must-read."

—Vivek Wadhwa, director of research at Duke University's Pratt School of Engineering and author of *The Immigrant Exodus*

THE SECOND MACHINE AGE

Also by Erik Brynjolfsson and Andrew McAfee

RACE AGAINST THE MACHINE

How the Digital Revolution Is Accelerating Innovation,
Driving Productivity, and Irreversibly Transforming Employment
and the Economy

Also by Erik Brynjolfsson

WIRED FOR INNOVATION

Also by Andrew McAfee

ENTERPRISE 2.0

New Collaborative Tools for your Organization's Toughest Challenges

ERIK BRYNJOLFSSON
ANDREW McAFEE

W. W. NORTON & COMPANY
NEW YORK LONDON

THE SECOND MACHINE AGE

Work, Progress, and Prosperity in a Time of Brilliant Technologies

Printed in the United States of America
First Edition

For information about special discounts for bulk purchases,
please contact W. W. Norton Special Sales at
specialsales@wwnorton.com or 800-233-4830

Manufacturing by Courier Westford
Book design by Lovedog Studio
Production manager: Devon Zahn

ISBN 978-0-393-23935-5

W. W. Norton & Company, Inc.
500 Fifth Avenue, New York, N.Y. 10110
www.wwnorton.com

W. W. Norton & Company Ltd.
Castle House, 75/76 Wells Street, London W1T 3QT

4 5 6 7 8 9 0

To Martha Pavlakis, the love of my life.

To my parents, David McAfee and Nancy Haller,
who prepared me for the second machine age by giving me
every advantage a person could have.

CONTENTS

THE SECOND MACHINE AGE

CHAPTER 1

THE BIG STORIES

"Technology is a gift of God. After the gift of life it is perhaps the greatest of God's gifts. It is the mother of civilizations, of arts and of sciences."

—Freeman Dyson

What have been the most important developments in human history?

As anyone investigating this question soon learns, it's difficult to answer. For one thing, when does 'human history' even begin? Anatomically and behaviorally modern *Homo sapiens*, equipped with language, fanned out from their African homeland some sixty thousand years ago.[1] By 25,000 BCE[2] they had wiped out the Neanderthals and other hominids, and thereafter faced no competition from other big-brained, upright-walking species.

We might consider 25,000 BCE a reasonable time to start tracking the big stories of humankind, were it not for the development-retarding ice age earth was experiencing at the time.[3] In his book *Why the West Rules—For Now*, anthropologist Ian Morris starts tracking human societal progress in 14,000 BCE, when the world clearly started getting warmer.

Another reason it's a hard question to answer is that it's not clear what criteria we should use: what constitutes a truly important development? Most of us share a sense that it would be an event or advance that significantly changes the course of things—one that 'bends the curve' of human history. Many have argued that the domestication of animals did just this, and is one of our earliest important achievements.

The dog might well have been domesticated before 14,000 BCE, but the horse was not; eight thousand more years would pass before we started breeding them and keeping them in corrals. The ox, too, had been tamed by that time (ca. 6,000 BCE) and hitched to a plow. Domestication of work animals hastened the transition from foraging to farming, an important development already underway by 8,000 BCE.[4]

Agriculture ensures plentiful and reliable food sources, which in turn enable larger human settlements and, eventually, cities. Cities in turn make tempting targets for plunder and conquest. A list of important human developments should therefore include great wars and the empires they yielded. The Mongol, Roman, Arab, and Ottoman empires—to name just four—were transformative; they affected kingdoms, commerce, and customs over immense areas.

Of course, some important developments have nothing to do with animals, plants, or fighting men; some are simply ideas. Philosopher Karl Jaspers notes that Buddha (563–483 BCE), Confucius (551–479 BCE), and Socrates (469–399 BCE) all lived quite close to one another in time (but not in place). In his analysis these men are the central thinkers of an 'Axial Age' spanning 800–200 BCE. Jaspers calls this age "a deep breath bringing the most lucid consciousness" and holds that its philosophers brought transformative schools of thought to three major civilizations: Indian, Chinese, and European.[5]

The Buddha also founded one of the world's major religions, and common sense demands that any list of major human developments include the establishment of other major faiths like Hinduism, Judaism, Christianity, and Islam. Each has influenced the lives and ideals of hundreds of millions of people.[6]

Many of these religions' ideas and revelations were spread by the written word, itself a fundamental innovation in human history. Debate rages about precisely when, where, and how writing was invented, but a safe estimate puts it in Mesopotamia around 3,200 BCE. Written symbols to facilitate counting also existed then, but they did not include the concept of zero, as basic as that seems to us now. The modern numbering system, which we call Arabic, arrived around 830 CE.[7]

The list of important developments goes on and on. The Athenians began to practice democracy around 500 BCE. The Black Death reduced Europe's population by at least 30 percent during the latter half of the 1300s. Columbus sailed the ocean blue in 1492, beginning interactions between the New World and the Old that would transform both.

The History of Humanity in One Graph

How can we ever get clarity about which of these developments is the *most* important? All of the candidates listed above have passionate advocates—people who argue forcefully and persuasively for one development's sovereignty over all the others. And in *Why the West Rules—For Now* Morris confronts a more fundamental debate: whether any attempt to rank or compare human events and developments is meaningful or legitimate. Many anthropologists and other social scientists say it is not. Morris disagrees, and his book boldly attempts to quantify human development. As he writes, "reducing the ocean of facts to simple numerical scores has drawbacks but it also has the one great merit of forcing everyone to confront the same evidence—with surprising results."[8] In other words, if we want to know which developments bent the curve of human history, it makes sense to try to draw that curve.

Morris has done thoughtful and careful work to quantify what he terms *social development* ("a group's ability to master its physical and intellectual environment to get things done") over time.* As Morris

* Morris defines human social development as consisting of four attributes: energy capture (per-person calories obtained from the environment for food, home and commerce, industry and agriculture, and transporta-

suggests, the results are surprising. In fact, they're astonishing. They show that none of the developments discussed so far has mattered very much, at least in comparison to something else—something that bent the curve of human history like nothing before or since. Here's the graph, with total worldwide human population graphed over time along with social development; as you can see, the two lines are nearly identical:

FIGURE 1.1 Numerically Speaking, Most of Human History Is Boring.

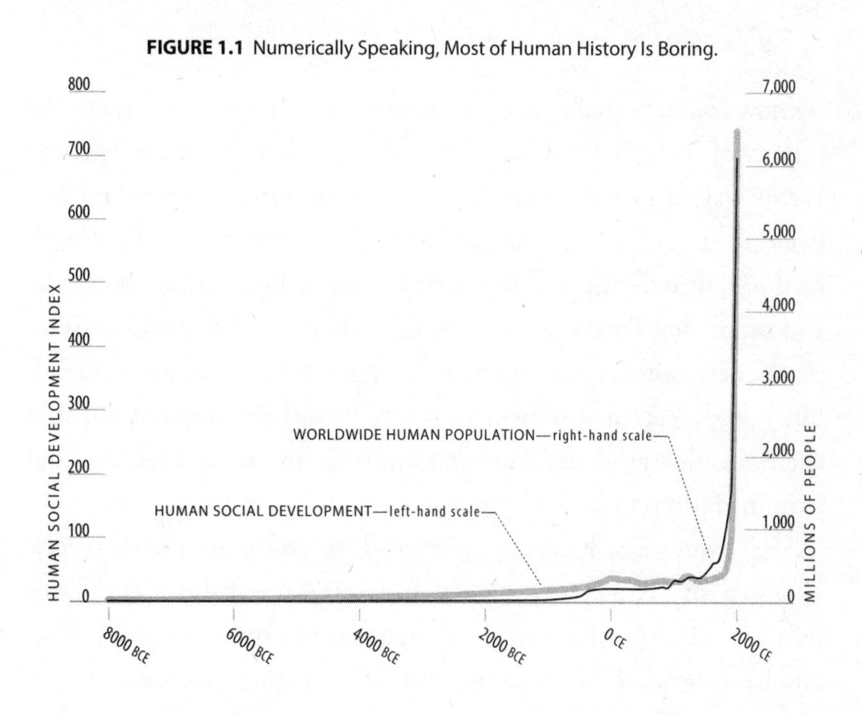

---

tion), organization (the size of the largest city), war-making capacity (number of troops, power and speed of weapons, logistical capabilities, and other similar factors), and information technology (the sophistication of available tools for sharing and processing information, and the extent of their use). Each of these is converted into a number that varies over time from zero to 250. Overall social development is simply the sum of these four numbers. Because he was interested in comparisons between the West (Europe, Mesopotamia, and North America at various times, depending on which was most advanced) and the East (China and Japan), he calculated social development separately for each area from 14,000 BCE to 2000 CE. In 2000, the East was higher only in organization (since Tokyo was the world's largest city) and had a social development score of 564.83. The West's score in 2000 was 906.37. We average the two scores.

For many thousands of years, humanity was a very gradual upward trajectory. Progress was achingly slow, almost invisible. Animals and farms, wars and empires, philosophies and religions all failed to exert much influence. But just over two hundred years ago, something sudden and profound arrived and bent the curve of human history—of population and social development—almost ninety degrees.

Engines of Progress

By now you've probably guessed what it was. This is a book about the impact of technology, after all, so it's a safe bet that we're opening it this way in order to demonstrate how important technology has been. And the sudden change in the graph in the late eighteenth century corresponds to a development we've heard a lot about: the Industrial Revolution, which was the sum of several nearly simultaneous developments in mechanical engineering, chemistry, metallurgy, and other disciplines. So you've most likely figured out that these technological developments underlie the sudden, sharp, and sustained jump in human progress.

If so, your guess is exactly right. And we can be even more precise about *which* technology was most important. It was the steam engine or, to be more precise, one developed and improved by James Watt and his colleagues in the second half of the eighteenth century.

Prior to Watt, steam engines were highly inefficient, harnessing only about one percent of the energy released by burning coal. Watt's brilliant tinkering between 1765 and 1776 increased this more than threefold.[9] As Morris writes, this made all the difference: "Even though [the steam] revolution took several decades to unfold . . . it was nonetheless the biggest and fastest transformation in the entire history of the world."[10]

The Industrial Revolution, of course, is not only the story of steam power, but steam started it all. More than anything else, it allowed us to overcome the limitations of muscle power, human and animal, and generate massive amounts of useful energy at will. This led to factories and mass production, to railways and mass transportation. It led, in other words, to modern life. The Industrial Revolution ushered in humanity's first machine age—the first time our progress was driven primarily by technological innovation—and it was the most profound time of transformation our world has ever seen.* The ability to generate massive amounts of mechanical power was so important that, in Morris's words, it "made mockery of all the drama of the world's earlier history."[11]

FIGURE 1.2 What Bent the Curve of Human History? The Industrial Revolution.

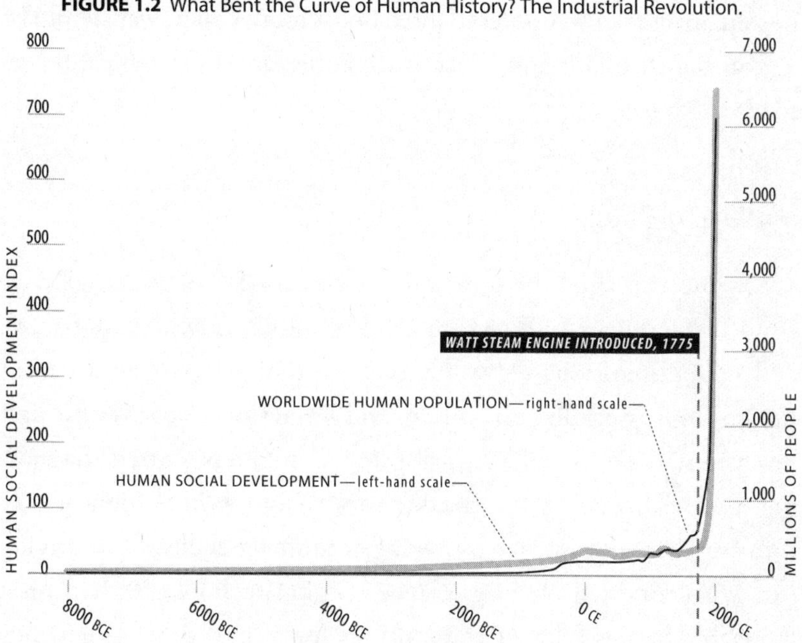

Now comes the second machine age. Computers and other digital advances are doing for mental power—the ability to use our brains

* We refer to the Industrial Revolution as the first machine age. However, "the machine age" is also a label used by some economic historians to refer to a period of rapid technological progress spanning the late nineteenth and early twentieth centuries. This same period is called by others the Second Industrial Revolution, which is how we'll refer to it in later chapters.

to understand and shape our environments—what the steam engine and its descendants did for muscle power. They're allowing us to blow past previous limitations and taking us into new territory. How exactly this transition will play out remains unknown, but whether or not the new machine age bends the curve as dramatically as Watt's steam engine, it is a very big deal indeed. This book explains how and why.

For now, a very short and simple answer: mental power is at least as important for progress and development—for mastering our physical and intellectual environment to get things done—as physical power. So a vast and unprecedented boost to mental power should be a great boost to humanity, just as the ealier boost to physical power so clearly was.

Playing Catch-Up

We wrote this book because we got confused. For years we have studied the impact of digital technologies like computers, software, and communications networks, and we thought we had a decent understanding of their capabilities and limitations. But over the past few years, they started surprising us. Computers started diagnosing diseases, listening and speaking to us, and writing high-quality prose, while robots started scurrying around warehouses and driving cars with minimal or no guidance. Digital technologies had been laughably bad at a lot of these things for a long time—then they suddenly got very good. How did this happen? And what were the implications of this progress, which was astonishing and yet came to be considered a matter of course?

We decided to team up and see if we could answer these questions. We did the normal things business academics do: read lots of papers and books, looked at many different kinds of data, and batted around ideas and hypotheses with each other. This was necessary and valuable, but the real learning, and the real fun, started when we went out into the world. We spoke with inventors, investors, entrepreneurs, engineers, scientists, and many others who make technology and put it to work.

Thanks to their openness and generosity, we had some futuristic experiences in today's incredible environment of digital innovation. We've ridden in a driverless car, watched a computer beat teams of Harvard and MIT students in a game of *Jeopardy!*, trained an industrial robot by grabbing its wrist and guiding it through a series of steps, handled a beautiful metal bowl that was made in a 3D printer, and had countless other mind-melting encounters with technology.

Where We Are

This work led us to three broad conclusions.

The first is that we're living in a time of astonishing progress with digital technologies—those that have computer hardware, software, and networks at their core. These technologies are not brand-new; businesses have been buying computers for more than half a century, and *Time* magazine declared the personal computer its "Machine of the Year" in 1982. But just as it took generations to improve the steam engine to the point that it could power the Industrial Revolution, it's also taken time to refine our digital engines.

We'll show why and how the full force of these technologies has recently been achieved and give examples of its power. "Full," though, doesn't mean "mature." Computers are going to continue to improve and to do new and unprecedented things. By "full force," we mean simply that the key building blocks are already in place for digital technologies to be as important and transformational to society and the economy as the steam engine. In short, we're at an inflection point—a point where the curve starts to bend a lot—because of computers. We are entering a second machine age.

Our second conclusion is that the transformations brought about by digital technology will be profoundly beneficial ones. We're heading into an era that won't just be different; it will be better, because we'll be able to increase both the variety and the volume of our consumption. When we phrase it that way—in the dry vocabulary of economics—it almost sounds unappealing. Who wants to consume more and more all the time? But we don't just consume calories and gasoline. We also consume information from books and friends, entertainment from superstars and amateurs, expertise from teachers and doctors, and countless other things that are not made of atoms. Technology can bring us more choice and even freedom.

When these things are digitized—when they're converted into bits that can be stored on a computer and sent over a network—they acquire some weird and wonderful properties. They're subject to different economics, where abundance is the norm rather than scarcity. As we'll show, digital goods are not like physical ones, and these differences matter.

Of course, physical goods are still essential, and most of us would like them to have greater volume, variety, and quality. Whether or not we want to eat more, we'd like to eat better or different meals. Whether or not we want to burn more fossil fuels, we'd like to visit more places with less hassle. Computers are helping accomplish these goals, and many others. Digitization is improving the physical world, and these improvements are only going to become more important. Among economic historians there's wide agreement that, as Martin Weitzman puts it, "the long-term growth of an advanced economy is dominated by the behavior of technical progress."[12] As we'll show, technical progress is improving exponentially.

Our third conclusion is less optimistic: digitization is going to bring with it some thorny challenges. This in itself should not be too surprising or alarming; even the most beneficial developments have unpleasant consequences that must be managed. The Industrial Revolution was accompanied by soot-filled London skies and horrific exploitation of child labor. What will be their modern equivalents? Rapid and accelerating digitization is likely to bring economic rather than environmental disruption, stemming from the fact that as computers get more powerful, companies have less need for some kinds of workers. Technological progress is going to leave behind some people, perhaps even a lot of people, as it races ahead. As we'll demonstrate, there's never been a better time to be a worker with special skills or the right education, because these people can use technology to create and capture value. However, there's never been a worse time to be a worker with only 'ordinary' skills and abilities to offer, because computers, robots, and other digital technologies are acquiring these skills and abilities at an extraordinary rate.

Over time, the people of England and other countries concluded that some aspects of the Industrial Revolution were unacceptable and took steps to end them (democratic government and technological progress both helped with this). Child labor no longer exists in the UK, and London air contains less smoke and sulfur dioxide now than at any time since at least the late 1500s.[13] The challenges of the digital revolution can also be met, but first we have to be clear on what they are. It's important to discuss the likely negative consequences of the second machine age and start a dialogue about how to mitigate them—we are confident that they're not insurmountable. But they won't fix themselves, either. We'll offer our thoughts on this important topic in the chapters to come.

So this is a book about the second machine age unfolding right now—an inflection point in the history of our economies and societies because of digitization. It's an inflection point in the right direction—bounty instead of scarcity, freedom instead of constraint—but one that will bring with it some difficult challenges and choices.

This book is divided into three sections. The first, composed of chapters 1 through 6, describes the fundamental characteristics of the second machine age. These chapters give many examples of recent technological progress that seem like the stuff of science fiction, explain why they're happening now (after all, we've had computers for decades), and reveal why we should be confident that the scale and pace of innovation in computers, robots, and other digital gear is only going to accelerate in the future.

The second part, consisting of chapters 7 through 11, explores bounty and spread, the two economic consequences of this progress. Bounty is the increase in volume, variety, and quality and the decrease in cost of the many offerings brought on by modern technological progress. It's the best economic news in the world today. Spread, however, is not so great; it's ever-bigger differences among people in economic success—in wealth, income, mobility, and other important measures. Spread has been increasing in recent years. This is a troubling development for many reasons, and one that will accelerate in the second machine age unless we intervene.

The final section—chapters 12 through 15—discusses what interventions will be appropriate and effective for this age. Our economic goals should be to maximize the bounty while mitigating the negative effects of the spread. We'll offer our ideas about how to best accomplish these aims, both in the short term and in the more distant future, when progress really has brought us into a world so technologically advanced that it seems to be the stuff of science fiction. As we stress in our concluding chapter, the choices we make from now on will determine what kind of world that is.

CHAPTER 2

THE SKILLS OF THE NEW MACHINES: TECHNOLOGY RACES AHEAD

"Any sufficiently advanced technology is indistinguishable from magic."

—Arthur C. Clarke

IN THE SUMMER OF 2012, we went for a drive in a car that had no driver.

During a research visit to Google's Silicon Valley headquarters, we got to ride in one of the company's autonomous vehicles, developed as part of its Chauffeur project. Initially we had visions of cruising in the back seat of a car that had no one in the front seat, but Google is understandably skittish about putting obviously autonomous autos on the road. Doing so might freak out pedestrians and other drivers, or attract the attention of the police. So we sat in the back while two members of the Chauffeur team rode up front.

When one of the Googlers hit the button that switched the car into fully automatic driving mode while we were headed down Highway 101, our curiosities—and self-preservation instincts—engaged. The 101 is not always a predictable or calm environment. It's nice and straight, but it's also crowded most of the time, and its traffic flows have little obvious rhyme or reason. At highway speeds the consequences of driving mistakes can be serious ones. Since we were now part of the ongoing Chauffeur experiment, these consequences were suddenly of more than just intellectual interest to us.

The car performed flawlessly. In fact, it actually provided a boring ride. It didn't speed or slalom among the other cars; it drove exactly the way we're all taught to in driver's ed. A laptop in the car provided a real-time visual representation of what the Google car 'saw' as it proceeded along the highway—all the nearby objects of which its sensors were aware. The car recognized all the surrounding vehicles, not just the nearest ones, and it remained aware of them no matter where they moved. It was a car without blind spots. But the software doing the driving was aware that cars and trucks driven by humans *do* have blind spots. The laptop screen displayed the software's best guess about where all these blind spots were and worked to stay out of them.

We were staring at the screen, paying no attention to the actual road, when traffic ahead of us came to a complete stop. The autonomous car braked smoothly in response, coming to a stop a safe distance behind the car in front, and started moving again once the rest of the traffic did. All the while the Googlers in the front seat never stopped their conversation or showed any nervousness, or indeed much interest at all in current highway conditions. Their hundreds of hours in the car had convinced them that it could handle a little stop-and-go traffic. By the time we pulled back into the parking lot, we shared their confidence.

The New *New Division of Labor*

Our ride that day on the 101 was especially weird for us because, only a few years earlier, we were sure that computers would not be able to drive cars. Excellent research and analysis, conducted by colleagues who we respect a great deal, concluded that driving would remain a human task for the foreseeable future. How they reached this conclusion, and how technologies like Chauffeur started to overturn it in just a few years, offers important lessons about digital progress.

In 2004 Frank Levy and Richard Murnane published their book *The New Division of Labor*.[1] The division they focused on was between human and digital labor—in other words, between people and computers. In any sensible economic system, people should focus on the tasks and jobs where they have a comparative advantage over computers, leaving computers the work for which they are better suited.

In their book Levy and Murnane offered a way to think about which tasks fell into each category.

One hundred years ago the previous paragraph wouldn't have made any sense. Back then, computers *were* humans. The word was originally a job title, not a label for a type of machine. Computers in the early twentieth century were people, usually women, who spent all day doing arithmetic and tabulating the results. Over the course of decades, innovators designed machines that could take over more and more of this work; they were first mechanical, then electro-mechanical, and eventually digital. Today, few people if any are employed simply to do arithmetic and record the results. Even in the lowest-wage countries there are no human computers, because the nonhuman ones are far cheaper, faster, and more accurate.

If you examine their inner workings, you realize that computers aren't just number crunchers, they're symbols processors. Their circuitry can be interpreted in the language of ones and zeroes, but equally validly as true or false, yes or no, or any other symbolic system. In principle, they can do all manner of symbolic work, from math to logic to language. But digital novelists are not yet available, so people still write all the books that appear on fiction bestseller lists. We also haven't yet computerized the work of entrepreneurs, CEOs, scientists, nurses, restaurant busboys, or many other types of workers. Why not? What is it about their work that makes it harder to digitize than what human computers used to do?

Computers Are Good at Following Rules . . .

These are the questions Levy and Murnane tackled in *The New Division of Labor*, and the answers they came up with made a great deal of sense. The authors put information processing tasks—the foundation of all knowledge work—on a spectrum. At one end are tasks like arithmetic that require only the application of well-understood rules. Since computers are really good at following rules, it follows that they should do arithmetic and similar tasks.

Levy and Murnane go on to highlight other types of knowledge work that can also be expressed as rules. For example, a person's credit score is a good general predictor of whether they'll pay back their mortgage as promised, as is the amount of the mortgage relative to the person's wealth, income, and other debts. So the decision about whether or not to give someone a mortgage can be effectively boiled down to a rule.

Expressed in words, a mortgage rule might say, "If a person is requesting a mortgage of amount *M* and they have a credit score of *V* or higher, annual income greater than *I* or total wealth greater than *W*, and total debt no greater than *D*, then approve the request." When expressed in computer code, we call a mortgage rule like this an *algorithm*. Algorithms are simplifications; they can't and don't take everything into account (like a billionaire uncle who has included the applicant in his will and likes to rock-climb without ropes). Algorithms do, however, include the most common and important things, and they generally work quite well at tasks like predicting payback rates. Computers, therefore, can and should be used for mortgage approval.*

. . . But Lousy at Pattern Recognition

At the other end of Levy and Murnane's spectrum, however, lie information processing tasks that cannot be boiled down to rules or algorithms. According to the authors, these are tasks that draw on

* In the years leading up to the Great Recession that began in 2007, companies were giving mortgages to people with lower and lower credit scores, income, and wealth, and higher and higher debt levels. In other words, they either rewrote or ignored their previous mortgage approval algorithms. It wasn't that the old mortgage algorithms stopped working; it was that they stopped being used.

the human capacity for pattern recognition. Our brains are extraordinarily good at taking in information via our senses and examining it for patterns, but we're quite bad at describing or figuring out *how* we're doing it, especially when a large volume of fast-changing information arrives at a rapid pace. As the philosopher Michael Polanyi famously observed, "We know more than we can tell."[2] When this is the case, according to Levy and Murnane, tasks can't be computerized and will remain in the domain of human workers. The authors cite driving a vehicle in traffic as an example of such as task. As they write,

> As the driver makes his left turn against traffic, he confronts a wall of images and sounds generated by oncoming cars, traffic lights, storefronts, billboards, trees, and a traffic policeman. Using his knowledge, he must estimate the size and position of each of these objects and the likelihood that they pose a hazard. . . . The truck driver [has] the schema to recognize what [he is] confronting. But articulating this knowledge and embedding it in software for all but highly structured situations are at present enormously difficult tasks. . . . Computers cannot easily substitute for humans in [jobs like driving].

So Much for *That* Distinction

We were convinced by Levy and Murnane's arguments when we read *The New Division of Labor* in 2004. We were further convinced that year by the initial results of the DARPA Grand Challenge for driverless cars.

DARPA, the Defense Advanced Research Projects Agency, was founded in 1958 (in response to the Soviet Union's launch of the *Sputnik* satellite) and tasked with spurring technological progress that might have military applications. In 2002 the agency announced its first Grand Challenge, which was to build a completely autonomous vehicle that could complete a 150-mile course through California's Mojave Desert. Fifteen entrants performed well enough in a qualifying run to compete in the main event, which was held on March 13, 2004.

The results were less than encouraging. Two vehicles didn't make it to the starting area, one flipped over *in* the starting area, and three hours into the race only four cars were still operational. The "winning" Sandstorm car from Carnegie Mellon University covered 7.4 miles (less than 5 percent of the total) before veering off the course during a hairpin turn and getting stuck on an embankment. The contest's $1 million prize went unclaimed, and *Popular Science* called the event "DARPA's Debacle in the Desert."[3]

Within a few years, however, the debacle in the desert became the 'fun on the 101' that we experienced. Google announced in an October 2010 blog post that its completely autonomous cars had for some time been driving successfully, in traffic, on American roads and highways. By the time we took our ride in the summer of 2012 the Chauffeur project had grown into a small fleet of vehicles that had collectively logged hundreds of thousands of miles with no human involvement and with only two accidents. One occurred when a person was driving the Chauffeur car; the other happened when a Google car was rear-ended (by a human driver) while stopped at a red light.[4] To be sure, there are still many situations that Google's cars can't handle, particularly complicated city traffic or off-road driving or, for that matter, any location that has not already been meticulously mapped in advance by Google. But our experience on the highway convinced us that it's a viable approach for the large and growing set of everyday driving situations.

Self-driving cars went from being the stuff of science fiction to on-the-road reality in a few short years. Cutting-edge research explaining why they were not coming anytime soon was outpaced by cutting-edge science and engineering that brought them into existence, again in the space of a few short years. This science and engineering accelerated rapidly, going from a debacle to a triumph in a little more than half a decade.

Improvement in autonomous vehicles reminds us of Hemingway's quote about how a man goes broke: "Gradually and then suddenly."[5] And self-driving cars are not an anomaly; they're part of a broad, fascinating pattern. Progress on some of the oldest and toughest challenges associated with computers, robots, and other digital gear was gradual for a long time. Then in the past few years it became sudden; digital gear started racing ahead, accomplishing tasks it had always been lousy at and displaying skills it was not supposed to acquire anytime soon. Let's look at a few more examples of surprising recent technological progress.

Good Listeners and Smooth Talkers

In addition to pattern recognition, Levy and Murnane highlight *complex communication* as a domain that would stay on the human side in the new division of labor. They write that, "Conversations critical to effective teaching, managing, selling, and many other occupations require the transfer and interpretation of a broad range of information. In these cases, the possibility of exchanging information with a computer, rather than another human, is a long way off."[6]

In the fall of 2011, Apple introduced the iPhone 4S featuring "Siri," an intelligent personal assistant that worked via a natural-language user interface. In other words, people talked to it just as they would talk to another human being. The software underlying Siri, which originated at the California research institute SRI International and was purchased by Apple in 2010, listened to what iPhone users were saying to it, tried to identify what they wanted, then took action and reported back to them in a synthetic voice.

After Siri had been out for about eight months, Kyle Wagner of technology blog *Gizmodo* listed some of its most useful capabilities:

"You can ask about the scores of live games—'What's the score of the Giants game?'—or about individual player stats. You can also make OpenTable reservations, get Yelp scores, ask about what movies are playing at a local theater and then see a trailer. If you're busy and can't take a call, you can ask Siri to remind you to call the person back later. This is the kind of everyday task for which voice commands can actually be incredibly useful."[7]

The *Gizmodo* post ended with caution: "That actually sounds pretty cool. Just with the obvious Siri criterion: *If it actually works.*"[8] Upon its release, a lot of people found that Apple's intelligent personal assistant didn't work well. It didn't understand what they were saying, asked for repeated clarifications, gave strange or inaccurate answers, and put them off with responses like "I'm really sorry about this, but I can't take any requests right now. Please try again in a little while." Analyst Gene Munster catalogued questions with which Siri had trouble:

* ***Where is Elvis buried?*** Responded, "I can't answer that for you." It thought the person's name was Elvis Buried.
* ***When did the movie* Cinderella *come out?*** Responded with a movie theater search on Yelp.
* ***When is the next Halley's Comet?*** Responded, "You have no meetings matching Halley's."
* ***I want to go to Lake Superior.*** Responded with directions to the company Lake Superior X-Ray.[9]

Siri's sometimes bizarre and frustrating responses became well known, but the power of the technology is undeniable. It can come to your aid exactly when you need it. On the same trip that afforded us some time in an autonomous car, we saw this firsthand. After a meeting in San Francisco, we hopped in our rental car to drive down to Google's headquarters in Mountain View. We had a portable GPS device with us, but didn't plug it in and turn it on because we thought we knew how to get to our next destination.

We didn't, of course. Confronted with an Escherian maze of elevated highways, off-ramps, and surface streets, we drove around looking for an on-ramp while tensions mounted. Just when our meeting at Google, this book project, and our professional relationship seemed in serious jeopardy, Erik pulled out his phone and asked Siri for "directions to U.S. 101 South." The phone responded instantly and flawlessly: the screen turned into a map showing where we were and how to find the elusive on-ramp.

We could have pulled over, found the portable GPS and turned it on, typed in our destination, and waited for our routing, but we didn't want to exchange information that way. We wanted to speak a question and hear and see (because a map was involved) a reply. Siri provided exactly the natural language interaction we were looking for. A 2004 review of the previous half-century's research in automatic speech recognition (a critical part of natural language processing) opened with the admission that "Human-level speech recognition has proved to be an elusive goal," but less than a decade later major elements of that goal have been reached. Apple and other companies have made robust natural language processing technology available to hundreds of millions of people via their mobile phones.[10] As noted by Tom Mitchell, who heads the machine-learning department at Carnegie Mellon University: "We're at the beginning of a ten-year period where we're going to transition from computers that can't understand language to a point where computers can understand quite a bit about language."[11]

Digital Fluency: The Babel Fish Goes to Work

Natural language processing software is still far from perfect, and computers are not yet as good as people at complex communication, but they're getting better all the time. And in tasks like translation from one language to another, surprising developments are underway: while computers' communication abilities are not as deep as those of the average human being, they're much broader.

A person who speaks more than one language can usually translate between them with reasonable accuracy. Automatic translation services, on the other hand, are impressive but rarely error-free. Even if your French is rusty, you can probably do better than Google Translate with the sentence "Monty Python's 'Dirty Hungarian Phrasebook' sketch is one of their funniest ones." Google offered, "Sketch des Monty Python 'Phrasebook sale hongrois' est l'un des plus drôles les leurs." This conveys the main gist, but has serious grammatical problems.

There is less chance you could have made progress translating this sentence (or any other) into Hungarian, Arabic, Chinese, Russian, Norwegian, Malay, Yiddish, Swahili, Esperanto, or any of the other sixty-three languages besides French that are part of the Google Translate service. But Google will attempt a translation of text from any of these languages into any other, instantaneously and at no cost for anyone with Web access.[12] The Translate service's smartphone app lets users speak more than fifteen of these languages into the phone and, in response, will produce synthesized, translated speech in more than half of the fifteen. It's a safe bet that even the world's most multilingual person can't match this breadth.

For years instantaneous translation utilities have been the stuff of science fiction (most notably *The Hitchhiker's Guide to the Galaxy's* Babel Fish, a strange creature that once inserted in the ear allows a person to understand speech in any language).[13] Google Translate and similar services are making it a reality today. In fact, at least one such service is being used right now to facilitate international customer service interactions. The translation services company Lionbridge has partnered with IBM to offer GeoFluent, an online application that instantly translates chats between customers and troubleshooters who do not share a language. In an initial trial, approximately 90 percent of GeoFluent users reported that it was good enough for business purposes.[14]

Human Superiority in *Jeopardy!*

Computers are now combining pattern matching with complex communication to quite literally beat people at their own games. In 2011, the February 14 and 15 episodes of the TV game show *Jeopardy!* included a contestant that was not a human being. It was a supercomputer called Watson, developed by IBM specifically to play the game (and named in honor of legendary IBM CEO Thomas Watson, Sr.). *Jeopardy!* debuted in 1964 and in 2012 was the fifth most popular syndicated TV program in America.[15] On a typical day almost 7 million people watch host Alex Trebek ask trivia questions on various topics as contestants vie to be the first to answer them correctly.*

The show's longevity and popularity stem from its being easy to understand yet extremely hard to play well. Almost everyone knows the answers to some of the questions in a given episode, but very few people know the answers to almost all of them. Questions cover a wide range of topics, and contestants are not told in advance what those topics will be. Players also have to be simultaneously fast, bold, and accurate—fast because they compete against one another for the chance to answer each question; bold because they have to try to answer a lot of questions, especially harder ones, in order to accumulate enough money to win; and accurate because money is subtracted for each incorrect answer.

Jeopardy!'s producers further challenge contestants with puns, rhymes, and other kinds of wordplay. A clue might ask, for example, for "A rhyming reminder of the past in the city of the NBA's Kings."[16]

* To be precise, Trebek reads answers and the contestants have to state the question that would give rise to this answer.

To answer correctly, a player would have to know what the acronym NBA stood for (in this case, it's the National Basketball Association, not the National Bank Act or chemical compound n-Butylamine), which city the NBA's Kings play in (Sacramento), and that the clue's demand for a *rhyming* reminder of the past meant that the right answer is "What is a Sacramento memento?" instead of a "Sacramento souvenir" or any other factually correct response. Responding correctly to clues like these requires mastery of pattern matching and complex communication. And winning at *Jeopardy!* requires doing both things repeatedly, accurately, and almost instantaneously.

During the 2011 shows, Watson competed against Ken Jennings and Brad Rutter, two of the best knowledge workers in this esoteric industry. Jennings won *Jeopardy!* a record seventy-four times in a row in 2004, taking home more than $3,170,000 in prize money and becoming something of a folk hero along the way.[17] In fact, Jennings is sometimes given credit for the existence of Watson.[18] According to one story circulating within IBM, Charles Lickel, a research manager at the company interested in pushing the frontiers of artificial intelligence, was having dinner in a steakhouse in Fishkill, New York, one night in the fall of 2004. At 7 p.m., he noticed that many of his fellow diners got up and went into the adjacent bar. When he followed them to find out what was going on, he saw that they were clustered in front of the bar's TV watching Jennings extend his winning streak beyond fifty matches. Lickel saw that a match between Jennings and a *Jeopardy!*-playing supercomputer would be extremely popular, in addition to being a stern test of a computer's pattern matching and complex communication abilities.

Since *Jeopardy!* is a three-way contest, the ideal third contestant would be Brad Rutter, who beat Jennings in the show's 2005 Ultimate Tournament of Champions and won more than $3,400,000.[19] Both men had packed their brains with information of all kinds, were deeply familiar with the game and all of its idiosyncrasies, and knew how to handle pressure.

These two humans would be tough for any machine to beat, and the first versions of Watson weren't even close. Watson could be 'tuned' by its programmers to be either more aggressive in answering questions (and hence more likely to be wrong) or more conservative and accurate. In December 2006, shortly after the project started, when Watson was tuned to try to answer 70 percent of the time (a relatively aggressive approach) it was only able to come up with the right response approximately 15 percent of the time. Jennings, in sharp contrast, answered about 90 percent of questions correctly in games when he buzzed in first (in other words, won the right to respond) 70 percent of the time.[20]

But Watson turned out to be a very quick learner. The supercomputer's performance on the aggression vs. accuracy tradeoff improved quickly, and by November 2010, when it was aggressive enough to win the right to answer 70 percent of a simulated match's total questions, it answered about 85 percent of them correctly. This was impressive improvement, but it still didn't put the computer in the same league as the best human players. The Watson team kept working until mid-January of 2011, when the matches were recorded for broadcast in February, but no one knew how well their creation would do against Jennings and Rutter.

Watson trounced them both. It correctly answered questions on topics ranging from "Olympic Oddities" (responding "pentathlon" to "A 1976 entry in the 'modern' this was kicked out for wiring his epee to score points without touching his foe") to "Church and State" (realizing that the answers all contained one or the other of these words, the computer answered "gestate" when told "It can mean to develop gradually in the mind or to carry during pregnancy"). While the supercomputer was not perfect (for example, it answered "chic" instead of "class" when asked about "stylish elegance, or students who all graduated in the same year" as part of the category "Alternate Meanings"), it was very good.

Watson was also extremely fast, repeatedly buzzing in before Jennings and Rutter to win the right to answer questions. In the first of the two games played, for example, Watson buzzed in first 43 times, then answered correctly 38 times. Jennings and Rutter *combined* to buzz in only 33 times over the course of the same game.[21]

At the end of the two-day tournament, Watson had amassed $77,147, more than three times as much as either of its human opponents. Jennings, who came in second, added a personal note on his answer to the tournament's final question: "I for one welcome our new computer overlords." He later elaborated, "Just as factory jobs were eliminated in the twentieth century by new assembly-line robots, Brad and I were the first knowledge-industry workers put out of work by the new generation of 'thinking' machines. 'Quiz show contestant' may be the first job made redundant by Watson, but I'm sure it won't be the last."[22]

The Paradox of Robotic 'Progress'

A final important area where we see a rapid recent acceleration in digital improvement is robotics—building machines that can navigate through and interact with the physical world of factories, warehouses, battlefields, and offices. Here again we see progress that was very gradual, then sudden.

The word *robot* entered the English language via the 1921 Czech play, *R.U.R.* (Rossum's "Universal" Robots) by Karel Capek, and automatons have been an object of human fascination ever since.[23] During the Great Depression, magazine and newspaper stories speculated that robots would wage war, commit crimes, displace workers, and even beat boxer Jack Dempsey.[24] Isaac Asimov coined the term *robotics* in 1941 and provided ground rules for the young discipline the following year with his famous Three Laws of Robotics:

1. A robot may not injure a human being or, through inaction, allow a human being to come to harm.

2. A robot must obey the orders given to it by human beings, except where such orders would conflict with the First Law.

3. A robot must protect its own existence as long as such protection does not conflict with the First or Second Laws.[25]

Asimov's enormous influence on both science fiction and real-world robot-making has persisted for seventy years. But one of those two communities has raced far ahead of the other. Science fiction has given us the chatty and loyal R2-D2 and C-3PO, *Battlestar Galactica*'s ominous Cylons, the terrible Terminator, and endless varieties of androids, cyborgs, and replicants. Decades of robotics research, in contrast, gave us Honda's ASIMO, a humanoid robot best known for a spectacularly failed demo that showcased its inability to follow Asimov's third law. At a 2006 presentation to a live audience in Tokyo, ASIMO attempted to walk up a shallow flight of stairs that had been placed on the stage. On the third step, the robot's knees buckled and it fell over backward, smashing its faceplate on the floor.[26]

ASIMO has since recovered and demonstrated skills like walking up and down stairs, kicking a soccer ball, and dancing, but its shortcomings highlight a broad truth: a lot of the things humans find easy and natural to do in the physical world have been remarkably difficult for robots to master. As the roboticist Hans Moravec has observed, "It is comparatively easy to make computers exhibit adult-level performance on intelligence tests or playing checkers, and difficult or impossible to give them the skills of a one-year-old when it comes to perception and mobility."[27]

This situation has come to be known as Moravec's paradox, nicely summarized by Wikipedia as "the discovery by artificial intelligence and robotics researchers that, contrary to traditional assumptions, high-level reasoning requires very little computation, but low-level sensorimotor skills require enormous computational resources."[28]* Moravec's insight is broadly accurate, and important. As the cognitive scientist Steven Pinker puts it, "The main lesson of thirty-five years of AI research is that the hard problems are easy and the easy problems are hard. . . . As the new generation of intelligent devices appears, it will be the stock analysts and petrochemical engineers and parole board members who are in danger of being replaced by machines. The gardeners, receptionists, and cooks are secure in their jobs for decades to come."[29]

Pinker's point is that robotics experts have found it fiendishly difficult to build machines that match the skills of even the least-trained manual worker. iRobot's Roomba, for example, can't do everything a maid does; it just vacuums the floor. More than ten million Roombas have been sold, but none of them is going to straighten the magazines on a coffee table.

When it comes to work in the physical world, humans also have a huge flexibility advantage over machines. Automating a single activity, like soldering a wire onto a circuit board or fastening two parts together with screws, is pretty easy, but that task must remain constant over time and take place in a 'regular' environment. For example, the circuit board must show up in exactly the same orientation every time. Companies buy specialized machines for tasks like these, have their engineers program and test them, then add them to their assembly lines. Each time the task changes—each time the location of the screw holes move, for example—production must stop until the machinery is reprogrammed. Today's factories, especially large ones in high-wage countries, are highly automated, but they're not full of general-purpose robots. They're full of dedicated, specialized machinery that's expensive to buy, configure, and reconfigure.

* Sensorimotor skills are those that involve sensing the physical world and controlling the body to move through it.

Rethinking Factory Automation

Rodney Brooks, who co-founded iRobot, noticed something else about modern, highly automated factory floors: people are scarce, but they're not absent. And a lot of the work they do is repetitive and mindless. On a line that fills up jelly jars, for example, machines squirt a precise amount of jelly into each jar, screw on the top, and stick on the label, but a person places the empty jars on the conveyor belt to start the process. Why hasn't this step been automated? Because in this case the jars are delivered to the line twelve at a time in cardboard boxes that don't hold them firmly in place. This imprecision presents no problem to a person (who simply sees the jars in the box, grabs them, and puts them on the conveyor belt), but traditional industrial automation has great difficulty with jelly jars that don't show up in exactly the same place every time.

In 2008 Brooks founded a new company, Rethink Robotics, to pursue and build *un*traditional industrial automation: robots that can pick and place jelly jars and handle the countless other imprecise tasks currently done by people in today's factories. His ambition is to make some progress against Moravec's paradox. What's more, Brooks envisions creating robots that won't need to be programmed by high-paid engineers; instead, the machines can be taught to do a task (or retaught to do a new one) by shop floor workers, each of whom need less than an hour of training to learn how to instruct their new mechanical colleagues. Brooks's machines are cheap, too. At about $20,000, they're a small fraction of the cost of current industrial robots. We got a sneak peek at these potential paradox-busters shortly before Rethink's public unveiling of their first line of robots, named Baxter. Brooks invited us to the company's headquarters in Boston to see these automatons, and to see what they could do.

Baxter is instantly recognizable as a humanoid robot. It has two burly, jointed arms with claw-like grips for hands; a torso; and a head with an LCD face that swivels to 'look at' the nearest person. It doesn't have legs, though; Rethink sidestepped the enormous challenges of automatic locomotion by putting Baxter on wheels and having it rely on people to get from place to place. The company's analyses suggest that it can still do lots of useful work without the ability to move under his own power.

To train Baxter, you grab it by the wrist and guide the arm through the motions you want it to carry out. As you do this, the arm seems weightless; its motors are working so you don't have to. The robot also maintains safety; the two arms can't collide (the motors resist you if you try to make this happen) and they automatically slow down if Baxter senses a person within their range. These and many other design features make working with this automaton a natural, intuitive, and nonthreatening experience. When we first approached it, we were nervous about catching a robot arm to the face, but this apprehension faded quickly, replaced by curiosity.

Brooks showed us several Baxters at work in the company's demo area. They were blowing past Moravec's paradox—sensing and manipulating lots of different objects with 'hands' ranging from grips to suction cups. The robots aren't as fast or fluid as a well-trained human worker at full speed, but they might not need to be. Most conveyor belts and assembly lines do not operate at full human speed; they would tire people out if they did.

Baxter has a few obvious advantages over human workers. It can work all day every day without needing sleep, lunch, or coffee breaks. It also won't demand healthcare from its employer or add to the payroll tax burden. And it can do two completely unrelated things at once; its two arms are capable of operating independently.

Coming Soon to Assembly Lines, Warehouses, and Hallways Near You

After visiting Rethink and seeing Baxter in action, we understood why Texas Instruments Vice President Remi El-Ouazzane said in early 2012, "We have a firm belief that the robotics market is on the cusp of exploding." There's a lot of evidence to support his view. The volume and variety of robots in use at companies is expanding rapidly, and innovators and entrepreneurs have recently made deep inroads against Moravec's paradox.[30]

Kiva, another young Boston-area company, has taught its automatons to move around warehouses safely, quickly, and effectively. Kiva robots look like metal ottomans or squashed R2-D2s. They scuttle around buildings at about knee-height, staying out of the way of humans and one another. They're low to the ground so they can scoot underneath shelving units, lift them up, and bring them to human workers. After these workers grab the products they need, the robot whisks the shelf away and another shelf-bearing robot takes its place. Software tracks where all the products, shelves, robots, and people are in the warehouse, and orchestrates the continuous dance of the Kiva automatons. In March of 2012, Kiva was acquired by Amazon—a leader in advanced warehouse logistics—for more than $750 million in cash.[31]

Boston Dynamics, yet another New England startup, has tackled Moravec's paradox head-on. The company builds robots aimed at supporting American troops in the field by, among other things, carrying heavy loads over rough terrain. Its BigDog, which looks like a giant metal mastiff with long skinny legs, can go up steep hills, recover from slips on ice, and do other very dog-like things. Balancing a heavy load on four points while moving over an uneven landscape is a truly nasty engineering problem, but Boston Dynamics has been making good progress.

As a final example of recent robotic progress, consider the Double, which is about as different from the BigDog as possible. Instead of trotting through rough enemy terrain, the Double rolls over cubicle carpets and hospital hallways carrying an iPad. It's essentially an upside-down pendulum with motorized wheels at the bottom and a tablet at the top of a four- to five-foot stick. The Double provides telepresence—it lets the operator 'walk around' a distant building and see and hear what's going on. The camera, microphone, and screen of the iPad serve as the eyes, ears, and face of the operator, who sees and hears what the iPad sees and hears. The Double itself acts as the legs, transporting the whole assembly around in response to commands from the operator. Double Robotics calls it "the simplest, most elegant way to be somewhere else in the world without flying there." The first batch of Doubles, priced at $2,499, sold out soon after the technology was announced in the fall of 2012.[32]

The next round of robotic innovation might put the biggest dent in Moravec's paradox ever. In 2012 DARPA announced another Grand Challenge; instead of autonomous cars, this one was about automatons. The DARPA Robotics Challenge (DRC) combined tool use, mobility, sensing, telepresence, and many other long-standing challenges in the field. According to the website of the agency's Tactical Technology Office,

> The primary technical goal of the DRC is to develop ground robots capable of executing complex tasks in dangerous, degraded, human-engineered environments. Competitors in the DRC are expected to focus on robots that can use standard tools and equipment commonly available in human environments, ranging from hand tools to vehicles, with an emphasis on adaptability to tools with diverse specifications.[33]

With the DRC, DARPA is asking the robotics community to build and demonstrate high-functioning humanoid robots by the end of 2014. According to an initial specification supplied by the agency, they will have to be able to drive a utility vehicle, remove debris blocking an entryway, climb a ladder, close a valve, and replace a pump.[34] These seem like impossible requirements, but we've been assured by highly knowledgeable colleagues—ones competing in the DRC, in fact—that they'll be met. Many saw the 2004 Grand Challenge as instrumental in accelerating progress with autonomous vehicles. There's an excellent chance that the DRC will be similarly important at getting us past Moravec's paradox.

More Evidence That We're at an Inflection Point

Self-driving cars, *Jeopardy!* champion supercomputers, and a variety of useful robots have all appeared just in the past few years. And these innovations are not just lab demos; they're showing off their skills and abilities in the messy real world. They contribute to the impression that we're at an inflection point—a bend in the curve where many technologies that used to be found only in science fiction are becoming everyday reality. As many other examples show, this is an accurate impression.

On the *Star Trek* television series, devices called tricorders were used to scan and record three kinds of data: geological, meteorological, and medical. Today's consumer smartphones serve all these purposes; they can be put to work as seismographs, real-time weather radar maps, and heart- and breathing-rate monitors.[35] And, of course, they're not limited to these domains. They also work as media players, game platforms, reference works, cameras, and GPS devices. On *Star Trek*, tricorders and person-to-person communicators were separate devices, but in the real world the two have merged in the smartphone. They enable their users to simultaneously access and generate huge amounts of information as they move around. This opens up the opportunity for innovations that venture capitalist John Doerr calls "SoLoMo"—social, local, and mobile.[36]

Computers historically have been very bad at writing real prose. In recent times they have been able to generate grammatically correct but meaningless sentences, a state of affairs that's been mercilessly exploited by pranksters. In 2008, for example, the International Conference on Computer Science and Software Engineering accepted the paper "Towards the Simulation of E-commerce" and invited its author to chair a session. This paper was 'written' by SCIgen, a program from the MIT Computer Science and Artificial Intelligence Lab that "generates random Computer Science research papers." SCIgen's authors wrote that, "Our aim here is to maximize amusement, rather than coherence," and after reading the abstract of "Towards the Simulation of E-commerce" it's hard to argue with them:[37]

> Recent advances in cooperative technology and classical communication are based entirely on the assumption that the Internet and active networks are not in conflict with object-oriented languages. In fact, few information theorists would disagree with the visualization of DHTs that made refining and possibly simulating 8 bitarchitectures a reality, which embodies the compelling principles of electrical engineering.[38]

Recent developments make clear, though, that not all computer-generated prose is nonsensical. Forbes.com has contracted with the company Narrative Science to write the corporate earnings previews that appear on the website. These stories are all generated by algorithms without human involvement. And they're indistinguishable from what a human would write:

> Forbes Earning Preview: H.J. Heinz
>
> A quality first quarter earnings announcement could push shares of H.J. Heinz (HNZ) to a new 52-week high as the price is just 49 cents off the milestone heading into the company's earnings release on Wednesday, August 29, 2012.
>
> The Wall Street consensus is 80 cents per share, up 2.6 percent from a year ago when H.J reported earnings of 78 cents per share.
>
> The consensus estimate remains unchanged over the past

> month, but it has decreased from three months ago when it was 82 cents. Analysts are expecting earnings of $3.52 per share for the fiscal year. Analysts project revenue to fall 0.3 percent year-over-year to $2.84 billion for the quarter, after being $2.85 billion a year ago. For the year, revenue is projected to roll in at $11.82 billion.[39]

Even computer peripherals like printers are getting in on the act, demonstrating useful capabilities that seem straight out of science fiction. Instead of just putting ink on paper, they are making complicated three-dimensional parts out of plastic, metal, and other materials. 3D printing, also sometimes called "additive manufacturing," takes advantage of the way computer printers work: they deposit a very thin layer of material (ink, traditionally) on a base (paper) in a pattern determined by the computer.

Innovators reasoned that there is nothing stopping printers from depositing layers one on top of the other. And instead of ink, printers can also deposit materials like liquid plastic that gets cured into a solid by ultraviolet light. Each layer is very thin—somewhere around one-tenth of a millimeter—but over time a three-dimensional object takes shape. And because of the way it is built up, this shape can be quite complicated—it can have voids and tunnels in it, and even parts that move independently of one another. At the San Francisco headquarters of Autodesk, a leading design software company, we handled a working adjustable wrench that was printed as a single part, no assembly required.[40]

This wrench was a demonstration product made out of plastic, but 3D printing has expanded into metals as well. Autodesk CEO Carl Bass is part of the large and growing community of additive manufacturing hobbyists and tinkerers. During our tour of his company's gallery, a showcase of all the products and projects enabled by Autodesk software, he showed us a beautiful metal bowl he designed on a computer and had printed out. The bowl had an elaborate lattice pattern on its sides. Bass said that he'd asked friends of his who were experienced in working with metal—sculptors, ironworkers, welders, and so on—how the bowl was made. None of them could figure out how the lattice was produced. The answer was that a laser had built up each layer by fusing powdered metal.

3D printing today is not just for art projects like Bass's bowl. It's used by countless companies every day to make prototypes and model parts. It's also being used for final parts ranging from plastic vents and housings on NASA's next-generation Moon rover to a metal prosthetic jawbone for an eighty-three-year-old woman. In the near future, it might be used to print out replacement parts for faulty engines on the spot instead of maintaining stockpiles of them in inventory. Demonstration projects have even shown that the technique could be used to build concrete houses.[41]

Most of the innovations described in this chapter have occurred in just the past few years. They've taken place in areas where improvement had been frustratingly slow for a long time, and where the best thinking often led to the conclusion that it wouldn't speed up. But then digital progress became sudden after being gradual for so long. This happened in multiple areas, from artificial intelligence to self-driving cars to robotics.

How did this happen? Was it a fluke—a confluence of a number of lucky one-time advances? No, it was not. The digital progress we've seen recently is certainly impressive, but it's just a small indication of what's to come. It's the dawn of the second machine age. To understand why it's unfolding now, we need to understand the nature of technological progress in the era of digital hardware, software, and networks. In particular, we need to understand its three key characteristics: that it is *exponential*, *digital*, and *combinatorial*. The next three chapters will discuss each of these in turn.

CHAPTER 3

MOORE'S LAW AND THE SECOND HALF OF THE CHESSBOARD

"The greatest shortcoming of the human race is our inability to understand the exponential function."

—Albert A. Bartlett

ALTHOUGH HE'S COFOUNDER OF Intel, a major philanthropist, and recipient of the Presidential Medal of Freedom, Gordon Moore is best known for a prediction he made, almost as an aside, in a 1965 article. Moore, then working at Fairchild Semiconductor, wrote an article for *Electronics* magazine with the admirably direct title "Cramming More Components onto Integrated Circuits." At the time, circuits of this type—which combined many different kinds of electrical components onto a single chip made primarily of silicon—were less than a decade old, but Moore saw their potential. He wrote that, "Integrated circuits will lead to such wonders as home computers—or at least terminals connected to a central computer—automatic controls for automobiles, and personal portable communications equipment."[1]

The article's most famous forecast, however, and the one that has made Moore a household name, concerned the component cramming of the title:

> The complexity for minimum component costs has increased at a rate of roughly a factor of two per year. . . . Certainly over the short term this rate can be expected to continue, if not to increase. Over the longer term, the rate of increase is a bit more uncertain, although there is no reason to believe it will not remain nearly constant for at least ten years.[2]

This is the original statement of Moore's Law, and it's worth dwelling for a moment on its implications. "Complexity for minimum component costs" here essentially means the amount of integrated circuit computing power you could buy for one dollar. Moore observed that over the relatively brief history of his industry this amount had doubled each year: you could buy twice as much power per dollar in 1963 as you could in 1962, twice as much again in 1964, and twice as much again in 1965.

Moore predicted this state of affairs would continue, perhaps with some change to timing, for at least another ten years. This bold statement forecast circuits that would be more than five hundred times as powerful in 1975 as they were in 1965.*

As it turned out, however, Moore's biggest mistake was in being too conservative. His "law" has held up astonishingly well for over four decades, not just one, and has been true for digital progress in areas other than integrated circuits. It's worth noting that the time required for digital doubling remains a matter of dispute. In 1975 Moore revised his estimate upward from one year to two, and today it's common to use eighteen months as the doubling period for general computing power. Still, there's no dispute that Moore's Law has proved remarkably prescient for almost half a century.[3]

It's Not a Law: It's a Bunch of Good Ideas

Moore's Law is very different from the laws of physics that govern thermodynamics or Newtonian classical mechanics. Those laws describe how the universe works; they're true no matter what we do. Moore's Law, in contrast, is a statement about the work of the computer industry's engineers and scientists; it's an observation about how constant and successful their efforts have been. We simply don't see this kind of sustained success in other domains.

There was no period of time when cars got twice as fast or twice as fuel efficient every year or two for fifty years. Airplanes don't consis-

* Since $2^9 = 512$

tently have the ability to fly twice as far, or trains the ability to haul twice as much. Olympic runners and swimmers don't cut their times in half over a generation, let alone a couple of years.

So how has the computer industry kept up this amazing pace of improvement?

There are two main reasons. First, while transistors and the other elements of computing are constrained by the laws of physics just like cars, airplanes, and swimmers, the constraints in the digital world are much looser. They have to do with how many electrons per second can be put through a channel etched in an integrated circuit, or how fast beams of light can travel through fiber-optic cable. At some point digital progress bumps up against its constraints and Moore's Law must slow down, but it takes awhile. Henry Samueli, chief technology officer of chipmaker Broadcom Corporation, predicted in 2013 that "Moore's Law is coming to an end—in the next decade it will pretty much come to an end so we have 15 years or so."[4]

But smart people have been predicting the end of Moore's Law for a while now, and they've been proved wrong over and over again.[5] This is not because they misunderstood the physics involved, but because they underestimated the people working in the computer industry. The second reason that Moore's Law has held up so well for so long is what we might call 'brilliant tinkering'—finding engineering detours around the roadblocks thrown up by physics. When it became difficult to cram integrated circuits more tightly together, for example, chip makers instead layered them on top of one another, opening up a great deal of new real estate. When communications traffic threatened to outstrip the capacity even of fiber-optic cable, engineers developed wavelength division multiplexing (WDM), a technique for transmitting many beams of light of different wavelengths down the same single glass fiber at the same time. Over and over again brilliant tinkering has found ways to skirt the limitations imposed by physics. As Intel executive Mike Marberry puts it, "If you're only using the same technology, then in principle you run into limits. The truth is we've been modifying the technology every five or seven years for 40 years, and there's no end in sight for being able to do that."[6] This constant modification has made Moore's Law the central phenomenon of the computer age. Think of it as a steady drumbeat in the background of the economy.

Charting the Power of Constant Doubling

Once this doubling has been going on for some time, the later numbers overwhelm the earlier ones, making them appear irrelevant. To see this, let's look at a hypothetical example. Imagine that Erik gives Andy a tribble, the fuzzy creature with a high reproductive rate made famous in an episode of *Star Trek*. Every day each tribble gives birth to another tribble, so Andy's menagerie doubles in size each day. A geek would say in this case that the tribble family is experiencing *exponential* growth. That's because the mathematical expression for how many tribbles there are on day x is 2^{x-1}, where the $x - 1$ is referred to as an exponent. Exponential growth like this is fast growth; after two weeks Andy has more than sixteen thousand of the creatures. Here's a graph of how his tribble family grows over time:

FIGURE 3.1 Tribbles over Time: The Power of Constant Doubling

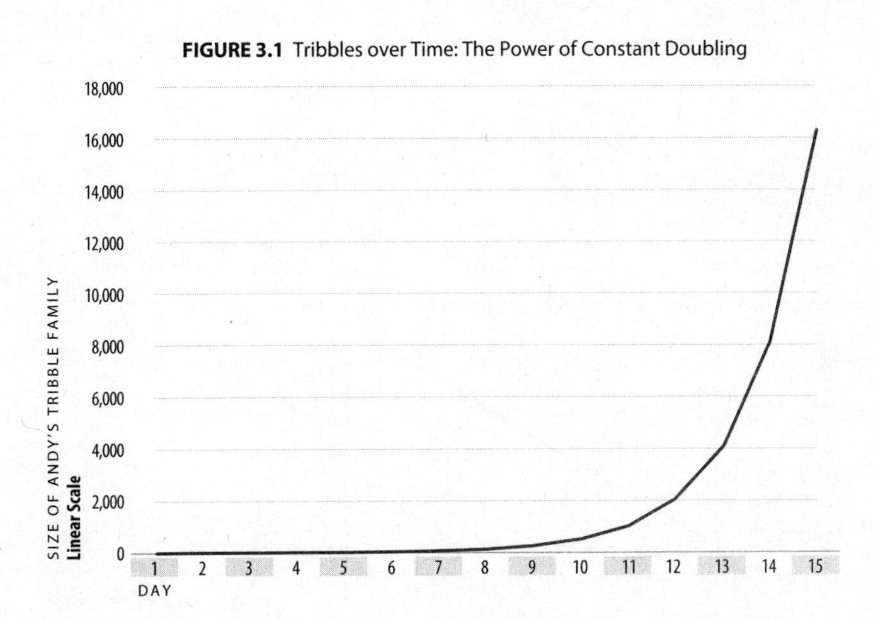

This graph is accurate, but misleading in an important sense. It seems to show that all the action occurs in the last couple of days, with nothing much happening in the first week. But the same phenomenon—the daily doubling of tribbles—has been going on the whole time with no accelerations or disruptions. This steady exponential growth is what's really interesting about Erik's 'gift' to Andy. To make it more obvious, we have to change the spacing of the numbers on the graph.

The graph we've already drawn has standard linear spacing; each segment of the vertical axis indicates two thousand more tribbles. This is fine for many purposes but, as we've seen, it's not great for showing exponential growth. To highlight it better, we'll change to logarithmic spacing, where each segment of the vertical axis represents a tenfold increase in tribbles: an increase first from 1 to 10, then from 10 to 100, then from 100 to 1,000, and so on. In other words, we scale the axis by powers of 10 or orders of magnitude.

Logarithmic graphs have a wonderful property: they show exponential growth as a perfectly straight line. Here's what the growth of Andy's tribble family looks like on a logarithmic scale:

FIGURE 3.2 Tribbles over Time: The Power of Constant Doubling

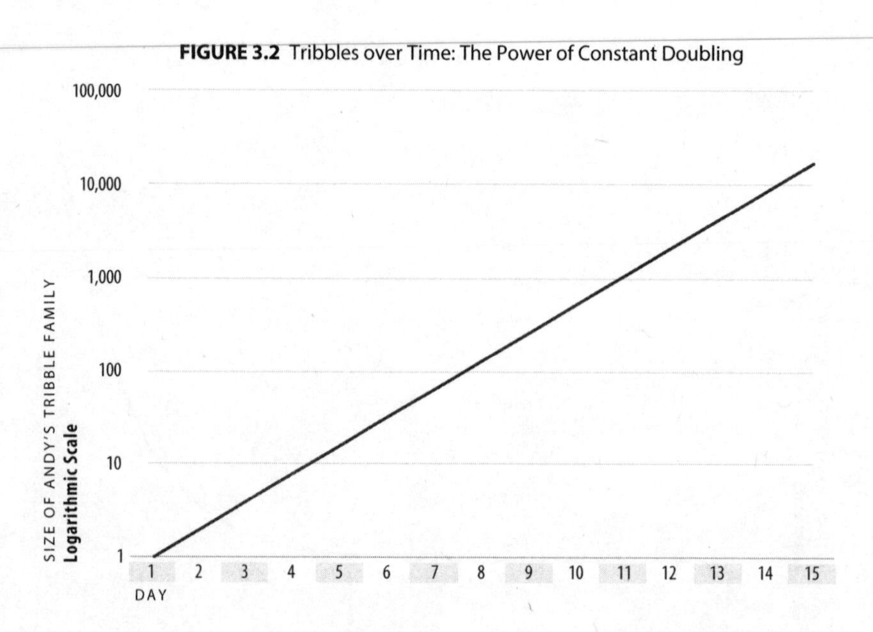

This view emphasizes the steadiness of the doubling over time rather than the large numbers at the end. Because of this, we often use logarithmic scales for graphing doublings and other exponential growth series. They show up as straight lines and their speed is easier to evaluate; the bigger the exponent, the faster they grow, and the steeper the line.

Impoverished Emperors, Headless Inventors, and the Second Half of the Chessboard

Our brains are not well equipped to understand sustained exponential growth. In particular, we severely underestimate how big the numbers can get. Inventor and futurist Ray Kurzweil retells an old story to drive this point home. The game of chess originated in present-day India during the sixth century CE, the time of the Gupta Empire.[7] As the story goes, it was invented by a very clever man who traveled to Pataliputra, the capital city, and presented his brainchild to the emperor. The ruler was so impressed by the difficult, beautiful game that he invited the inventor to name his reward.

The inventor praised the emperor's generosity and said, "All I desire is some rice to feed my family." Since the emperor's largess was spurred by the invention of chess, the inventor suggested they use the chessboard to determine the amount of rice he would be given. "Place one single grain of rice on the first square of the board, two on the second, four on the third, and so on," the inventor proposed, "so that each square receives twice as many grains as the previous."

"Make it so," the emperor replied, impressed by the inventor's apparent modesty.

Moore's Law and the tribble exercise allow us to see what the emperor did not: sixty-three instances of doubling yields a fantastically big number, even when starting with a single unit. If his request were fully honored, the inventor would wind up with 2^{64-1}, or more than eighteen quintillion grains of rice. A pile of rice this big would dwarf Mount Everest; it's more rice than has been produced in the history of the world. Of course, the emperor could not honor such a request. In some versions of the story, once he realizes that he's been tricked, he has the inventor beheaded.

Kurzweil tells the story of the inventor and the emperor in his 2000 book *The Age of Spiritual Machines: When Computers Exceed Human Intelligence.* He aims not only to illustrate the power of sustained exponential growth but also to highlight the point at which the numbers start to become so big they are inconceivable:

> After thirty-two squares, the emperor had given the inventor about 4 billion grains of rice. That's a reasonable quantity—about one large field's worth—and the emperor did start to take notice.
>
> But the emperor could still remain an emperor. And the inventor could still retain his head. It was as they headed into the second half of the chessboard that at least one of them got into trouble.[8]

Kurzweil's great insight is that while numbers do get large in the first half of the chessboard, we still come across them in the real world. Four billion does not necessarily outstrip our intuition. We experience it when harvesting grain, assessing the fortunes of the world's richest people today, or tallying up national debt levels. In the second half of the chessboard, however—as numbers mount into trillions, quadrillions, and quintillions—we lose all sense of them. We also lose sense of how quickly numbers like these appear as exponential growth continues.

Kurzweil's distinction between the first and second halves of the chessboard inspired a quick calculation. Among many other things, the U.S. Bureau of Economic Analysis (BEA) tracks American companies' expenditures. The BEA first noted "information technology" as a distinct corporate investment category in 1958. We took that year as the starting point for when Moore's Law entered the business world, and used eighteen months as the doubling period. After thirty-two of these doublings, U.S. businesses entered the second half of the chessboard when it comes to the use of digital gear. That was in 2006.

Of course, this calculation is just a fun little exercise, not anything like a serious attempt to identify the one point at which everything changed in the world of corporate computing. You could easily argue with the starting point of 1958 and a doubling period of eighteen months. Changes to either assumption would yield a different break point between the first and second halves of the chessboard. And business technologists were not only innovating in the second half; as we'll discuss later, the breakthroughs of today and tomorrow rely on, and would be impossible without, those of the past.

We present this calculation here because it underscores an important idea: that exponential growth eventually leads to staggeringly big numbers, ones that leave our intuition and experience behind. In other words, things get weird in the second half of the chessboard. And like the emperor, most of us have trouble keeping up.

One of the things that sets the second machine age apart is how quickly that second half of the chessboard can arrive. We're not claiming that no other technology has ever improved exponentially. In fact, after the one-time burst of improvement in the steam engine Watt's innovations created, additional tinkering led to exponential improvement over the ensuing two hundred years. But the exponents were relatively small, so it only went through about three or four doublings in efficiency during that period.[9] It would take a millennium to reach the second half of the chessboard at that rate. In the second machine age, the doublings happen much faster and exponential growth is much more salient.

Second-Half Technologies

Our quick doubling calculation also helps us understand why progress with digital technologies feels so much faster these days and why we've seen so many recent examples of science fiction becoming business reality. It's because the steady and rapid exponential growth of Moore's Law has added up to the point that we're now in a different regime of computing: we're now in the second half of the chessboard. The innovations we described in the previous chapter—cars that drive themselves in traffic; *Jeopardy!*-champion supercomputers; auto-generated news stories; cheap, flexible factory robots; and inexpensive consumer devices that are simultaneously communicators, tricorders, and computers—have all appeared since 2006, as have countless other marvels that seem quite different from what came before.

One of the reasons they're all appearing now is that the digital gear at their hearts is finally both fast and cheap enough to enable them. This wasn't the case just a decade ago. What does digital progress look like on a logarithmic scale? Let's take a look.

FIGURE 3.3 The Many Dimensions of Moore's Law

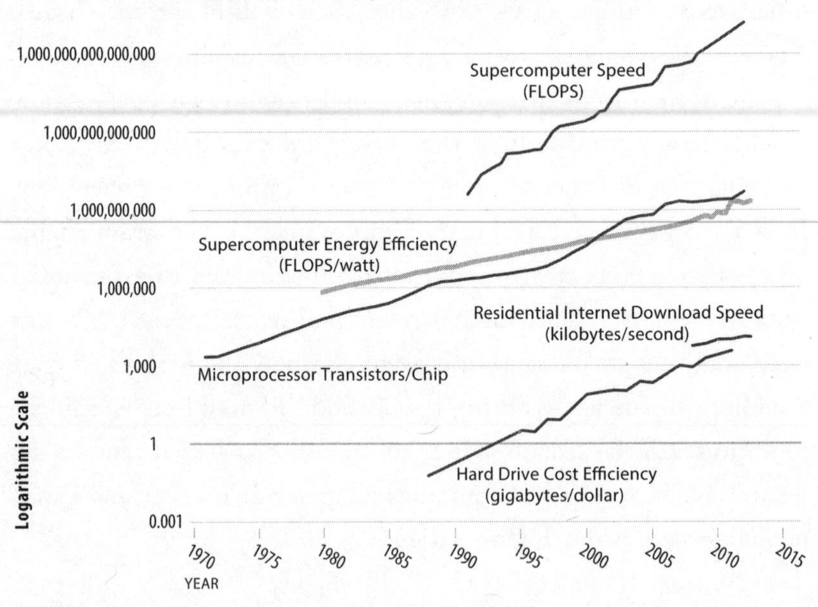

This graph shows that Moore's Law is both consistent and broad; it's been in force for a long time (decades, in some cases) and applies to many types of digital progress. As you look at it, keep in mind that if it used standard linear scaling on the vertical axis, all of those straight-ish lines would look like the first graph above of Andy's tribble family—horizontal most of the way, then suddenly close to vertical at the end. And there would really be no way to graph them all together—the numbers involved are just too different. Logarithmic scaling takes care of these issues and allows us to get a clear overall picture of improvement in digital gear.

It's clear that many of the critical building blocks of computing—microchip density, processing speed, storage capacity, energy efficiency, download speed, and so on—have been improving at exponential rates for a long time. To understand the real-world impacts of Moore's Law, let's compare the capabilities of computers separated by only a few doubling periods. The ASCI Red, the first product of the U.S. government's Accelerated Strategic Computing Initiative, was the world's fastest supercomputer when it was introduced in 1996. It cost $55 million to develop and its one hundred cabinets occupied nearly 1,600 square feet of floor space (80 percent of a tennis court) at Sandia National Laboratories in New Mexico.[10] Designed for calculation-intensive tasks like simulating nuclear tests, ASCI Red was the first computer to score above one teraflop—one trillion floating point operations* per second—on the standard benchmark test for computer speed. To reach this speed it used eight hundred kilowatts per hour, about as much as eight hundred homes would. By 1997, it had reached 1.8 teraflops.

Nine years later another computer hit 1.8 teraflops. But instead of simulating nuclear explosions, it was devoted to drawing them and other complex graphics in all their realistic, real-time, three-dimensional glory. It did this not for physicists, but for video game players. This computer was the Sony PlayStation 3, which matched the ASCI Red in performance, yet cost about five hundred dollars,

* Multiplying 62.34 by 24358.9274 is an example of a floating point operation. The decimal point in such operations is allowed to 'float' instead of being fixed in the same place for both numbers.

took up less than a tenth of a square meter, and drew about two hundred watts.[11] In less than ten years exponential digital progress brought teraflop calculating power from a single government lab to living rooms and college dorms all around the world. The PlayStation 3 sold approximately 64 million units. The ASCI Red was taken out of service in 2006.

Exponential progress has made possible many of the advances discussed in the previous chapter. IBM's Watson draws on a plethora of clever algorithms, but it would be uncompetitive without computer hardware that is about one hundred times more powerful than Deep Blue, its chess-playing predecessor that beat the human world champion, Garry Kasparov, in a 1997 match. Speech recognition applications like Siri require lots of computing power, which became available on mobile phones like Apple's iPhone 4S (the first phone that came with Siri installed). The iPhone 4S was about as powerful, in fact, as Apple's top-of-the-line Powerbook G4 laptop had been a decade earlier. As all of these innovations show, exponential progress allows technology to keep racing ahead and makes science fiction reality in the second half of the chessboard.

Not Just for Computers Anymore: The Spread of Moore's Law

Another comparison across computer generations highlights not only the power of Moore's Law but also its wide reach. As is the case with the ASCI Red and the PlayStation 3, the Cray-2 supercomputer (introduced in 1985) and iPad 2 tablet (introduced in 2011) had almost identical peak calculation speeds. But the iPad also had a speaker, microphone, and headphone jack. It had two cameras; the one on the front of the device was Video Graphics Array (VGA) quality, while the one on the back could capture high-definition video. Both could also take still photographs, and the back camera had a 5x digital zoom. The tablet had receivers that allowed it to participate in both wireless telephone and Wi-Fi networks. It also had a GPS receiver, digital compass, accelerometer, gyroscope, and light sensor. It had no built-in keyboard, relying instead on a high-definition touch screen that could track up to eleven points of contact simultaneously.[12] It fit all of this capability into a device that cost much less than $1,000 and was smaller, thinner, and lighter than many magazines. The Cray-2, which cost more than $35 million (in 2011 dollars), was by comparison deaf, dumb, blind, and immobile.[13]

Apple was able to cram all of this functionality in the iPad 2 because a broad shift has taken place in recent decades: sensors like microphones, cameras, and accelerometers have moved from the analog world to the digital one. They became, in essence, computer chips. As they did so, they became subject to the exponential improvement trajectories of Moore's Law.

Digital gear for recording sounds was in use by the 1960s, and an Eastman Kodak engineer built the first modern digital camera in 1975.[14] Early devices were expensive and clunky, but quality quickly improved and prices dropped. Kodak's first digital single-lens reflex camera, the DCS 100, cost about $13,000 when it was introduced in 1991; it had a maximum resolution of 1.3 megapixels and stored its images in a separate, ten-pound hard drive that users slung over their shoulders. However, the pixels per dollar available from digital cameras doubled about every year (a phenomenon known as "Hendy's Law" after Kodak Australia employee Barry Hendy, who documented it), and all related gear got exponentially smaller, lighter, cheaper, and better over time.[15] Accumulated improvement in digital sensors meant that twenty years after the DCS 100, Apple could include two tiny cameras, capable of both still and video photography, on the iPad 2. And when it introduced a new iPad the following year, the rear camera's resolution had improved by a factor of more than seven.

Machine Eyes

As Moore's Law works over time on processors, memory, sensors, and many other elements of computer hardware (a notable exception is batteries, which haven't improved their performance at an exponential rate because they're essentially chemical devices, not digital ones), it does more than just make computing devices faster, cheaper, smaller, and lighter. It also allows them to do things that previously seemed out of reach.

Researchers in artificial intelligence have long been fascinated (some would say obsessed) with the problem of simultaneous localization and mapping, which they refer to as SLAM. SLAM is the process of building up a map of an unfamiliar building as you're navigating through it—where are the doors? where are stairs? what are all the things I might trip over?—and also keeping track of where you are within it (so you can find your way back downstairs and out the front door). For the great majority of humans, SLAM happens with minimal conscious thought. But teaching machines to do it has been a huge challenge.

Researchers thought a great deal about which sensors to give a robot (cameras? lasers? sonar?) and how to interpret the reams of data they provide, but progress was slow. As a 2008 review of the topic summarized, SLAM "is one of the fundamental challenges of robotics . . . [but it] seems that almost all the current approaches can not perform consistent maps for large areas, mainly due to the increase of the computational cost and due to the uncertainties that become prohibitive when the scenario becomes larger."[16] In short, sensing a sizable area and immediately crunching all the resulting data were thorny problems preventing real progress with SLAM. Until, that is, a $150 video-game accessory came along just two years after the sentences above were published.

In November 2010 Microsoft first offered the Kinect sensing

device as an addition to its Xbox gaming platform. The Kinect could keep track of two active players, monitoring as many as twenty joints on each. If one player moved in front of the other, the device made a best guess about the obscured person's movements, then seamlessly picked up all joints once he or she came back into view. Kinect could also recognize players' faces, voices, and gestures and do so across a wide range of lighting and noise conditions. It accomplished this with digital sensors including a microphone array (which pinpointed the source of sound better than a single microphone could), a standard video camera, and a depth perception system that both projected and detected infrared light. Several onboard processors and a great deal of proprietary software converted the output of these sensors into information that game designers could use.[17] At launch, all of this capability was packed into a four-inch-tall device less than a foot wide that retailed for $149.99.

The Kinect sold more than eight million units in the sixty days after its release (more than either the iPhone or iPad) and currently holds the Guinness World Record for the fastest-selling consumer electronics device of all time.[18] The initial family of Kinect-specific games let players play darts, exercise, brawl in the streets, and cast spells à la Harry Potter.[19] These, however, did not come close to exhausting the system's possibilities. In August of 2011 at the SIGGRAPH (short for the Association of Computing Machinery's Special Interest Group on Graphics and Interactive Techniques) conference in Vancouver, British Columbia, a team of Microsoft employees and academics used Kinect to "SLAM" the door shut on a long-standing challenge in robotics.

SIGGRAPH is the largest and most prestigious gathering devoted to research and practice on digital graphics, attended by researchers, game designers, journalists, entrepreneurs, and most others interested in the field. This made it an appropriate place for Microsoft to unveil what the Creators Project website called "The Self-Hack That

Could Change Everything."*[20] This was the KinectFusion, a project that used the Kinect to tackle the SLAM problem.

In a video shown at SIGGRAPH 2011, a person picks up a Kinect and points it around a typical office containing chairs, a potted plant, and a desktop computer and monitor.[21] As he does, the video splits into multiple screens that show what the Kinect is able to sense. It immediately becomes clear that if the Kinect is not completely solving the SLAM problem for the room, it's coming close. In real time, Kinect draws a three-dimensional map of the room and all the objects in it, including a coworker. It picks up the word DELL pressed into the plastic on the back of the computer monitor, even though the letters are not colored and only one millimeter deeper that the rest of the monitor's surface. The device knows where it is in the room at all times, and even knows how virtual ping-pong balls would bounce around if they were dropped into the scene. As the technology blog *Engadget* put it in a post-SIGGRAPH entry, "The Kinect took 3D sensing to the mainstream, and moreover, allowed researchers to pick up a commodity product and go absolutely nuts."[22]

In June of 2011, shortly before SIGGRAPH, Microsoft had made available a Kinect software development kit (SDK) giving programmers everything they needed to start writing PC software that made use of the device. After the conference there was a great deal of interest in using the Kinect for SLAM, and many teams in robotics and AI research downloaded the SDK and went to work.

In less than a year, a team of Irish and American researchers led by our colleague John Leonard of MIT's Computer Science and Artificial Intelligence Lab announced Kintinuous, a "spatially extended" version of KinectFusion. With Kintinuous, users could use a Kinect to scan large indoor volumes like apartment buildings and even outdoor environments (which the team scanned by holding a Kinect

* In this context, a "hack" is an effort to get inside the guts of a piece of digital gear and use it for an unorthodox purpose. A self-hack is one carried out by the company that made the gear in the first place.

outside a car window during a nighttime drive). At the end of the paper describing their work, the Kintinuous researchers wrote, "In the future we will extend the system to implement a full SLAM approach."[23] We don't think it will be long until they announce success. When given to capable technologists, the exponential power of Moore's Law eventually makes even the toughest problems tractable.

Cheap and powerful digital sensors are essential components of some of the science-fiction technologies discussed in the previous chapter. The Baxter robot has multiple digital cameras and an array of force and position detectors. All of these would have been unworkably expensive, clunky, and imprecise just a short time ago. A Google autonomous car incorporates several sensing technologies, but its most important 'eye' is a Cyclopean LIDAR (a combination of "LIght" and "raDAR") assembly mounted on the roof. This rig, manufactured by Velodyne, contains sixty-four separate laser beams and an equal number of detectors, all mounted in a housing that rotates ten times a second. It generates about 1.3 million data points per second, which can be assembled by onboard computers into a real-time 3D picture extending one hundred meters in all directions. Some early commercial LIDAR systems available around the year 2000 cost up to $35 million, but in mid-2013 Velodyne's assembly for self-navigating vehicles was priced at approximately $80,000, a figure that will fall much further in the future. David Hall, the company's founder and CEO, estimates that mass production would allow his product's price to "drop to the level of a camera, a few hundred dollars."[24]

All these examples illustrate the first element of our three-part explanation of why we're now in the second machine age: steady exponential improvement has brought us into the second half of the chessboard—into a time when what's come before is no longer a particularly reliable guide to what will happen next. The accumulated doubling of Moore's Law, and the ample doubling still to come, gives us a world where supercomputer power becomes available to toys in just a few years, where ever-cheaper sensors enable inexpensive solutions to previously intractable problems, and where science fiction keeps becoming reality.

Sometimes a difference in degree (in other words, more of the same) becomes a difference in kind (in other words, different than anything else). The story of the second half of the chessboard alerts us that we should be aware that enough exponential progress can take us to astonishing places. Multiple recent examples convince us that we're already there.

CHAPTER 4

THE DIGITIZATION OF JUST ABOUT EVERYTHING

"When you can measure what you are speaking about, and express it in numbers, you know something about it; but when you cannot express it in numbers, your knowledge is of a meagre and unsatisfactory kind."

—Lord Kelvin

"HEY, HAVE YOU HEARD about . . . ?"

"You've got to check out . . . "

Questions and recommendations like these are the stuff of everyday life. They're how we learn about new things from our friends, family, and colleagues, and how we spread the word about exciting things we've come across. Traditionally, such cool hunting ended with the name of a band, restaurant, place to visit, TV show, book, or movie.

In the digital age, sentences like these frequently end with the name of a website or a gadget. And right now, they're often about a smartphone application. Both of the major technology platforms in this market—Apple's iOS and Google's Android—have more than five hundred thousand applications available.[1] There are plenty of "Top 10" and "Best of" lists available to help users find the cream of the smartphone app crop, but traditional word of mouth has retained its power.

Not long ago Matt Beane, a doctoral student at the MIT Sloan School of Management and a member of our Digital Frontier team, gave us a tip. "You've got to check out Waze; it's amazing." But when we found out it was a GPS-based app that provided driving directions, we weren't immediately impressed. Our cars have navigation systems and our iPhones can give driving directions through the Maps application. We could not see a need for yet another how-do-I-get-there technology.

As Matt patiently explained, using Waze is like bringing a Ducati to a drag race against an oxcart. Unlike traditional GPS navigation, Waze doesn't tell you what route to your destination is best in general; it tells you what route is best *right now.* As the company website explains:

> The idea for Waze originated years ago, when Ehud Shabtai . . . was given a PDA with an external GPS device pre-installed with navigation software. Ehud's initial excitement quickly gave way to disappointment—the product didn't reflect the dynamic changes that characterize real conditions on the road. . . .
>
> Ehud took matters into his own hands. . . . His goal? To accurately reflect the road system, state of traffic and all the information relevant to drivers at any given moment.[2]

Anyone who has used a traditional GPS system will recognize Shabtai's frustration. Yes, they know your precise location thanks to a network of twenty-four geosynchronous GPS satellites built and maintained by the U.S. government. They also know about roads—which ones are highways, one-way streets, and so on—because they have access to a database with this information. But that's about it. The things a driver really wants to know about—traffic jams, accidents, road closures, and other factors that affect travel time—escape a traditional system. When asked, for example, to calculate the best route from Andy's house to Erik's, it simply takes the starting point (Andy's car's current location) and the ending point (Erik's house) and consults its road database to calculate the theoretically "quickest" route between the two. This route will include major roads and highways, since they have the highest speed limits.

If it's rush hour, however, this theoretically quickest route will not actually be the quickest one; with thousands of cars squeezing onto the major roads and highways, traffic speed will not approach, let alone eclipse, the speed limit. Andy should instead seek out all the sneaky little back roads that longtime commuters know about. Andy's GPS knows that these roads exist (if it's up-to-date, it knows about *all* roads), but doesn't know that they're the best option at eight forty-five on a Tuesday morning. Even if he starts out on back roads, his well-meaning GPS will keep rerouting him onto the highway.

Shabtai recognized that a truly useful GPS system needed to know more than where the car was on the road. It also needed to know where *other* cars were and how fast they were moving. When the first smartphones appeared he saw an opportunity, founding Waze in 2008 along with Uri Levine and Amir Shinar. The software's genius is to turn all the smartphones running it into sensors that upload constantly to the company's servers their location and speed information. As more and more smartphones run the application, therefore, Waze gets a more and more complete sense of how traffic is flowing throughout a given area. Instead of just a static map of roads, it also has always current updates on traffic conditions. Its servers use the map, these updates, and a set of sophisticated algorithms to generate driving directions. If Andy wants to drive to Erik's at 8:45 a.m. on a Tuesday, Waze is not going to put him on the highway. It's going to keep him on surface streets where traffic is comparatively light at that hour.

That Waze gets more useful to all of its members as it gets more members is a classic example of what economists call a *network effect*—a situation where the value of a resource for each of its users increases with each additional user. And the number of Wazers, as they're called, is increasing quickly. In July of 2012 the company reported that it had doubled its user base to twenty million people in the previous six months.[3] This community had collectively driven more than 3.2 billion miles and had typed in many thousands of updates about accidents, sudden traffic jams, police speed traps, road closings, new freeway exits and entrances, cheap gas, and other items of interest to their fellow drivers.

Waze makes GPS what it should be for drivers: a system for getting where you want to go as quickly and easily as possible, regardless of how much you know about local roads and conditions. It instantly turns you into the most knowledgeable driver in town.

The Economics of Bits

Waze is possible in no small part because of Moore's Law and exponential technological progress, the subjects of the previous chapter. The service relies on vast numbers of powerful but cheap devices (the smartphones of its users), each of them equipped with an array of processors, sensors, and transmitters. Such technology simply didn't exist a decade ago, and so neither did Waze. It only became feasible in the past few years because of accumulated digital power increases and cost declines. As we saw in chapter 3, exponential improvement in computer gear is one of the three fundamental forces enabling the second machine age.

Waze also depends critically on the second of these three forces: digitization. In their landmark 1998 book *Information Rules*, economists Carl Shapiro and Hal Varian define this phenomenon as "encod[ing information] as a stream of bits."[4] Digitization, in other words, is the work of turning all kinds of information and media—text, sounds, photos, video, data from instruments and sensors, and so on—into the ones and zeroes that are the native language of computers and their kin. Waze, for example, uses several streams of information: digitized street maps, location coordinates for cars broadcast by the app, and alerts about traffic jams, among others. It's Waze's ability to bring these streams together and make them useful for its users that causes the service to be so popular.

We thought we understood digitization pretty well based on the work of Shapiro, Varian, and others, and based on our almost constant exposure to online content, but in the past few years the phenomenon has evolved in some unexpected directions. It has also exploded in volume, velocity, and variety. This surge in digitization has had two profound consequences: new ways of acquiring knowledge (in other words, of doing science) and higher rates of innovation. This chapter will explore the fascinating recent history of digitization.

Like so many other modern online services, Waze exploits two of the well-understood and unique economic properties of digital information: such information is *non-rival*, and it has *close to zero marginal cost of reproduction*. In everyday language, we might say that digital information is not "used up" when it gets used, and it is extremely cheap to make another copy of a digitized resource. Let's look at each of these properties in a bit more detail.

Rival goods, which we encounter every day, can only be consumed by one person or thing at a time. If the two of us fly from Boston to California, the plane that takes off after us cannot use our fuel. Andy can't also have the seat that Erik is sitting in (airline rules prohibit such sharing, even if we were up for it) and can't use his colleague's headphones if Erik has already put them on to listen to music on his smartphone. The digitized music itself, however, is non-rival. Erik's listening to it doesn't keep anyone else from doing so, at the same time or later.

If Andy buys and reads an old hardcover copy of the collected works of science-fiction writer Jules Verne, he doesn't "use it up"; he can pass it on to Erik once he's done. But if the two of us want to dip into *Twenty Thousand Leagues Under the Sea* at the same time, we either have to find another copy or Andy has to make a copy of the book he owns. He might be legally entitled to do this because it's not under copyright, but he'd still have to spend a lot of time at the photocopier or pay someone else to do so. In either case, making that copy would not be cheap.[5] In addition, a photocopy of a photocopy of a photocopy starts to get hard to read.

But if Andy has acquired a digital copy of the book, with a couple keystrokes or mouse clicks he can create a duplicate, save it to a physical disk, and give the copy to Erik. Unlike photocopies, bits cloned from bits are usually exactly identical to the original. Copying bits is also extremely cheap, fast, and easy to do. While the very first copy of a book or movie might cost a lot to create, making additional copies cost almost nothing. This is what is meant by "zero marginal cost of reproduction."

These days, of course, instead of handing Erik a disk, Andy is more likely to attach the file to an e-mail message or share it through a cloud service like Dropbox. One way or another, though, he's going to use the Internet. He'll take this approach because it's faster, more convenient, and, in an important sense, essentially free. Like most people, we pay a flat fee for Internet access at home and on our mobile devices (MIT pays for our access at work). If we exceed a certain data limit, our Internet Service Provider might start charging us extra, but until that point we don't pay by the bit; we pay the same no matter how many bits we upload or download. As such, there's no additional cost for sending or receiving one more chunk of data over the Net. Unlike goods made of atoms, goods made of bits can be replicated perfectly and sent across the room or across the planet almost instantaneously and almost costlessly. Making things free, perfect, and instant might seem like unreasonable expectations for most products, but as more information is digitized, more products will fall into these categories.

Business Models When the First Copy is Still Expensive

Shapiro and Varian elegantly summarize these attributes by stating that in an age of computers and networks, "Information is costly to produce but cheap to reproduce."[6] Instantaneous online translation services, one of the science-fiction-into-reality technologies discussed in chapter 2, take advantage of this fact. They make use of paired sets of documents that were translated, often at considerable expense, by a human from one language into another. For example, the European Union and its predecessor bodies have since 1957 issued all official documents in all the main languages of its member countries, and the United Nations has been similarly prolific in writing texts in all six of its official languages.

This huge body of information was not cheap to generate, but once it's digitized it's very cheap to replicate, chop up, and share widely and repeatedly. This is exactly what a service like Google Translate does. When it gets an English sentence and a request for its German equivalent, it essentially scans all the documents it knows about in both English and German, looking for a close match (or a few fragments that add up to a close match), then returns the corresponding German text. Today's most advanced automatic translation services, then, are not the result of any recent insight about how to teach computers all the rules of human languages and how to apply them. Instead, they're applications that do statistical pattern matching over huge pools of digital content that was costly to produce, but cheap to reproduce.

What Happens When the Content Comes Freely?

But what would happen to the digital world if information were no longer costly to produce? What would happen if it were free right from the start? We've been learning the answers to these questions in the years since *Information Rules* came out, and they're highly encouraging.

The old business saying is that "time is money," but what's amazing about the modern Internet is how many people are willing to devote their time to producing online content without seeking any money in return. Wikipedia's content, for example, is generated for free by volunteers all around the world. It's by far the world's largest and most consulted reference work, but no one gets paid to write or edit its articles. The same is true for countless websites, blogs, discussion boards, forums, and other sources of online information. Their creators expect no direct monetary reward and offer the information free of charge.

When Shapiro and Varian published *Information Rules* in 1998, the rise of such user-generated content, much of which is created without money changing hands, had yet to occur. Blogger, one of the first weblog services, debuted in August 1999, Wikipedia in January 2001, and Friendster, an early social networking site, in 2002. Friendster was soon eclipsed by Facebook, which was founded in 2004 and has since grown into the most popular Internet site in the world.[7] In fact, six of the ten most popular content sites throughout the world are primarily user-generated, as are six of the top ten in the United States.[8]

All this user-generated content isn't just making us feel good by letting us express ourselves and communicate with one another; it's also contributing to some of the recent science-fiction-into-reality technologies we've seen. Siri, for example, improves itself over time by analyzing the ever-larger collection of sound files its users generate when interacting with the voice recognition system. And Watson's database, which consisted of approximately two hundred million pages of documents taking up four terabytes of disk space, included an entire copy of Wikipedia.[9] For a while it also included the salty language–filled Urban Dictionary, but this archive of user-generated content was removed after, to the dismay of its creators, Watson started to include curse words in its responses.[10]

Perhaps we shouldn't be too surprised by the growth and popularity of user-generated content on the Internet. After all, we humans like to share and interact. What's a bit more surprising is how much our machines also apparently like talking to each other.

Machine-to-machine (M2M) communication is a catch-all term

for devices sharing data with one another over networks like the Internet. Waze makes use of M2M; when the app is active on a smartphone, it constantly sends information to Waze's servers without any human involvement. Similarly, when you search the popular travel site Kayak for cheap airfares, Kayak's servers immediately send requests to their counterparts at various airlines, which write back in real time without any human involvement. ATMs ask their banks how much money we have in our accounts before letting us withdraw cash; digital thermometers in refrigerated trucks constantly reassure supermarkets that the produce isn't getting too hot in transit; sensors in semiconductor factories let headquarters know every time a defect occurs; and countless other M2M communications take place in real time, all the time. According to a July 2012 story in the *New York Times*, "The combined level of robotic chatter on the world's wireless networks . . . is likely soon to exceed that generated by the sum of all human voice conversations taking place on wireless grids."[11]

Running Out of Metric System: The Data Explosion

The digitization of just about everything—documents, news, music, photos, video, maps, personal updates, social networks, requests for information and responses to those requests, data from all kinds of sensors, and so on—is one of the most important phenomena of recent years. As we move deeper into the second machine age, digitization continues to spread and accelerate, yielding some jaw-dropping statistics. According to Cisco Systems, worldwide Internet traffic increased by a factor of twelve in just the five years between 2006 and 2011, reaching 23.9 exabytes per month.[12]

An *exabyte* is a ridiculously big number, the equivalent of more than two hundred thousand of Watson's entire database. However, even this is not enough to capture the magnitude of current and future digitization. Technology research firm IDC estimates that there were 2.7 zettabytes, or 2.7 sextillion bytes, of digital data in the world in 2012, almost half as much again as existed in 2011. And this data won't just sit on disk drives; it'll also move around. Cisco predicts that global Internet Protocol traffic will reach 1.3 zettabytes by 2016.[13] That's over 250 billion DVDs of information.[14]

As these figures make clear, digitization yields truly big data. In fact, if this kind of growth keeps up for much longer we're going to run out of metric system. When its set of prefixes was expanded in 1991 at the nineteenth General Conference on Weights and Measures, the largest one was *yotta*, signifying one septillion, or 10^{24}.[15] We're only one prefix away from that in the 'zettabyte era.'

Binary Science

The recent explosion of digitization is clearly impressive, but is it important? Are all of these exa- and zettabytes of digital data actually useful?

They're incredibly useful. One of the main reasons we cite digitization as a main force shaping the second machine age is that digitization increases understanding. It does this by making huge amounts of data readily accessible, and data are the lifeblood of science. By "science" here, we mean the work of formulating theories and hypotheses, then evaluating them. Or, less formally, guessing how something works, then checking to see if the guess is right.

A while back Erik guessed that data about Internet searches might signal future changes in housing sales and prices around the country. He reasoned that if a couple is going to move to another city and buy a house, they are not going to complete the process in just a few days. They're going to start investigating the move and purchase months in advance. These days those initial investigations will take place over the Internet and consist of typing into a search engine phrases like "Phoenix real estate agent," "Phoenix neighborhoods," and "Phoenix two-bedroom house prices."

To test this hypothesis, Erik asked Google if he could access data about its search terms. He was told that he didn't have to ask; the company made these data freely available over the Web. Erik and his doctoral student Lynn Wu, neither of whom was versed in the economics of housing, built a simple statistical model to look at the data utilizing the user-generated content of search terms made available by Google. Their model linked changes in search-term volume to later housing sales and price changes, predicting that if search terms like the ones above were on the increase today, then housing sales and prices in Phoenix would rise three months from now. They found their simple model worked. In fact, it predicted sales 23.6 percent more accurately than predictions published by the experts at the National Association of Realtors.

Researchers have had similar success using newly available digital data in other domains. A team led by Rumi Chunara of Harvard Medical School found that tweets were just as accurate as official reports when it came to tracking the spread of cholera after the 2010 earthquake in Haiti; they were also at least two weeks faster.[16] Sitaram Asur and Bernardo Huberman of HP's Social Computing Lab found that tweets could also be used to predict movie box-office revenue. They concluded that "this work shows how social media expresses a collective wisdom which, when properly tapped, can yield an extremely powerful and accurate indicator of future outcomes."[17]

Digitization can also help us better understand the past. As of March 2012 Google had scanned more than twenty million books published over several centuries.[18] This huge pool of digital words and phrases forms a base for what's being called *culturomics*, or "the application of high-throughput data collection and analysis to the study of human culture."[19] A multidisciplinary team led by Jean-Baptiste Michel and Erez Lieberman Aiden analyzed over five million books published in English since 1800. Among other things, they found that the number of words in English increased by more than 70 percent between 1950 and 2000, that fame now comes to people more quickly than in the past but also fades faster, and that in the twentieth century interest in evolution was declining until Watson and Crick discovered the structure of DNA.[20]

All of these are examples of better understanding and prediction—in other words, of better science—via digitization. Hal Varian, who's now Google's chief economist, has for years enjoyed a front-row seat for this phenomenon. He also has a way with words. One of our favorite quotes of his is, "I keep saying that the sexy job in the next ten years will be statisticians. And I'm not kidding."[21] When we look at the amount of digital data being created and think about how much more insight there is to be gained, we're pretty sure he's not wrong, either.

New Layers Yield New Recipes

Digital information isn't just the lifeblood for new kinds of science; it's the second fundamental force (after exponential improvement) shaping the second machine age because of its role in fostering innovation. Waze is a great example here. The service is built on multiple layers and generations of digitization, none of which have decayed or been used up since digital goods are non-rival.

The first and oldest layer is digital maps, which are at least as old as personal computers.[22] The second is GPS location information, which became much more useful for driving when the U.S. government increased its GPS accuracy in 2000.[23] The third is social data; Waze users help each other by providing information on everything from accidents to police speed traps to cheap gas; they can even use the app to chat with one another. And finally, Waze makes extensive use of sensor data; in fact, it essentially converts every car using it into a traffic-speed sensor and uses these data to calculate the quickest routes.

In-car navigation systems that use only the first two generations of digital data—maps and GPS location information—have been around for a while. They can be extremely useful, especially in unfamiliar cities, but as we've seen, they have serious shortcomings. The founders of Waze realized that as digitization advanced and spread they could overcome the shortcomings of traditional GPS navigation. These innovators made progress by adding social and sensor data to an existing system, greatly increasing its power and usefulness. As we'll see in the next chapter, this style of innovation is one of the hallmarks of our current time. It's so important, in fact, that it's the third and last of the forces shaping the second machine age. The next chapter explains why this is.

CHAPTER 5

INNOVATION: DECLINING OR RECOMBINING?

"If you want to have good ideas you must have many ideas."

—Linus Pauling

Everyone agrees that it would be troubling news if America's rate of innovation were to decrease. But we can't seem to agree at all about whether this is actually happening.

We care about innovation so much not simply because we like new stuff, although we certainly do. As the novelist William Makepeace Thackeray observed, "Novelty has charms that our mind can hardly withstand."[1] Some of us can hardly withstand the allure of new gadgets; others are charmed by the latest fashion styles or places to see and be seen. From an economist's perspective, satisfying these desires is great—taking care of consumer demand is usually seen as a good thing. But innovation is also the most important force that makes our society wealthier.

Why Innovation is (Almost) Everything

Paul Krugman speaks for many, if not most, economists when he says, "Productivity isn't everything, but in the long run it is almost everything." Why? Because, he explains, "A country's ability to improve its standard of living over time depends almost entirely on its ability to raise its output per worker"—in other words, the number of hours of labor it takes to produce everything, from automobiles to zippers, that we produce.[2] Most countries don't have extensive mineral wealth or oil reserves, and thus can't get rich by exporting them.* So the only viable way for societies to become wealthier—to improve the standard of living available to its people—is for their companies and workers to keep getting more output from the same number of inputs, in other words more goods and services from the same number of people.

Innovation is how this productivity growth happens. Economists love to argue with one another, but there's great consensus among them about the fundamental importance of innovation for growth and prosperity. Most in the profession would agree with Joseph Schumpeter, the topic's great scholar, who wrote that, "Innovation is the outstanding fact in the economic history of capitalist society . . . and also it is largely responsible for most of what we would at first sight attribute to other factors."[3] It is here that the consensus ends. How much of this "outstanding fact" is taking place right now, and whether it's on an upward or downward trend, is a matter of great dispute.

Why We Should Be Worried: Innovations Get Used Up

Economist Bob Gordon, one of the most thoughtful, thorough, and widely respected researchers of productivity and economic growth, recently completed a major study of how the American standard of living has changed over the past 150 years. His work left him convinced that innovation is slowing down.

Gordon emphasizes—as do we—the role of new technologies in driving economic growth. And like us, he's impressed by the productive power unleashed by the steam engine and the other technologies of the Industrial Revolution. According to Gordon, it was

* In reality, many of the countries that *do* have large amounts of mineral and commodity wealth are often crippled by the twin terrors of the "resource curse": low growth rates and lots of poverty.

the first truly significant event in the economic history of the world. As he writes, "there was almost no economic growth for four centuries and probably for the previous millennium" prior to 1750, or roughly when the Industrial Revolution started.[4] As we saw in the first chapter, human population growth and social development were very nearly flat until the steam engine came along. Unsurprisingly, it turns out that economic growth was, too.

As Gordon shows, however, once this growth got started it stayed on a sharp upward trajectory for two hundred years. This was due not only to the original Industrial Revolution, but also to a second one, it too reliant on technological innovation. Three novelties were central here: electricity, the internal combustion engine, and indoor plumbing with running water, all of which came onto the scene between 1870 and 1900.

The 'great inventions' of this second industrial revolution, in Gordon's estimation, "were so important and far-reaching that they took a full 100 years to have their main effect." But once that effect had been realized, a new problem emerged. Growth stalled out, and even began to decline. At the risk of being flippant, when the steam engine ran out of steam, the internal combustion engine was there to replace it. But once the internal combustion engine ran out of fuel, we weren't left with much. To use Gordon's words,

> The growth of productivity (output per hour) slowed markedly after 1970. While puzzling at the time, it seems increasingly clear that the one-time-only benefits of the Great Inventions and their spin-offs had occurred and could not happen again. . . . All that remained after 1970 were second-round improvements, such as developing short-haul regional jets, extending the original interstate highway network with suburban ring roads, and converting residential America from window unit air conditioners to central air conditioning.[5]

Gordon is far from alone in this view. In his 2011 book *The Great Stagnation*, economist Tyler Cowen is definitive about the source of America's economic woes:

> We are failing to understand why we are failing. All of these problems have a single, little noticed root cause: We have been living off low-hanging fruit for at least three hundred years. . . . Yet during the last forty years, that low-hanging fruit started disappearing, and we started pretending it was still there. We have failed to recognize that we are at a technological plateau and the trees are more bare than we would like to think.[6]

General Purpose Technologies: The Ones That Really Matter

Clearly, Gordon and Cowen see the invention of powerful technologies as central to economic progress. Indeed, there's broad agreement among economic historians that some technologies are significant enough to accelerate the normal march of economic progress. To do this, they have to spread throughout many, if not most, industries; they can't remain in just one. The cotton gin, for example, was unquestionably important within the textile sector at the start of the nineteenth century, but pretty insignificant outside of it.*

The steam engine and electrical power, by contrast, quickly spread just about everywhere. The steam engine didn't just massively increase the amount of power available to factories and free them from the need to be located near a stream or river to power the water wheel; it also revolutionized land travel by enabling railroads and sea travel via the steamship. Electricity gave a further boost to manufac-

* Some have tied the invention of the cotton gin to increased demand for slave labor in the American South and therefore to the Civil War, but its direct economic effect outside the textile industry was minimal.

turing by enabling individually powered machines. It also lit factories, office buildings, and warehouses and led to further innovations like air conditioning, which made previously sweltering workplaces pleasant.

With their typical verbal flair, economists call innovations like steam power and electricity *general purpose technologies* (GPTs). Economic historian Gavin Wright offers a concise definition: "deep new ideas or techniques that have the potential for important impacts on many sectors of the economy."[7] "Impacts" here mean significant boosts to output due to large productivity gains. GPTs are important because they are economically significant—they interrupt and accelerate the normal march of economic progress.

In addition to agreeing on their importance, scholars have also come to a consensus on how to recognize GPTs: they should be pervasive, improving over time, and able to spawn new innovations.[8] The preceding chapters have built a case that digital technologies meet all three of these requirements. They improve along a Moore's Law trajectory, are used in every industry in the world, and lead to innovations like autonomous cars and nonhuman *Jeopardy!* champions. Are we alone in thinking that information and communication technology (ICT) belongs in the same category as steam and electricity? Are we the only ones who think, in short, that ICT is a GPT?

Absolutely not. Most economic historians concur with the assessment that ICT meets all of the criteria given above, and so should join the club of general purpose technologies. In fact, in a list of all the candidates for this classification compiled by the economist Alexander Field, only steam power got more votes than ICT, which was tied with electricity as the second most commonly accepted GPT.[9]

If we are all in agreement, then why the debate over whether ICTs are ushering in a new golden age of innovation and growth? Because, the argument goes, their economic benefits have already been captured and now most new 'innovation' involves entertaining ourselves inexpensively online. According to Robert Gordon:

> The first industrial robot was introduced by General Motors in 1961. Telephone operators went away in the 1960s. . . . Airline reservations systems came in the 1970s, and by 1980 barcode scanners and cash machines were spreading through the retail and banking industries. . . . The first personal computers arrived in the early 1980s with their word processing, word wrap, and spreadsheets. . . . More recent and thus more familiar was the rapid development of the web and e-commerce after 1995, a process largely completed by 2005.[10]

At present, says Cowen, "The gains of the Internet are very real and I am here to praise them, not damn them. . . . Still, the overall picture is this: We are having more fun, in part because of the Internet. We are also having more cheap fun. [But] we are coming up short on the revenue side, so it is harder to pay our debts, whether individuals, businesses, or governments."[11] Twenty-first century ICT, in short, is failing the prime test of being economically significant.

Why We Shouldn't Be Worried: Innovations Don't Get Used Up

For any good scientist, of course, data are the ultimate decider of hypotheses. So what do the data say here? Do the productivity numbers back up this pessimistic view of the power of digitization? We'll get to the data in chapter 7. First, though, we want to present a very different view of how innovation works—an alternative to the notion that innovations get 'used up.'

Gordon writes that "it is useful to think of the innovative process as a series of discrete inventions followed by incremental improvements which ultimately tap the full potential of the initial invention."[12]

This seems sensible enough. An invention like the steam engine or computer comes along and we reap economic benefits from it. Those benefits start small while the technology is immature and not widely used, grow to be quite big as the GPT improves and propagates, then taper off as the improvement—and especially the propagation—die down. When multiple GPTs appear at the same time, or in a steady sequence, we sustain high rates of growth over a long period. But if there's a big gap between major innovations, economic growth will eventually peter out. We'll call this the 'innovation-as-fruit' view of things, in honor of Tyler Cowen's imagery of all the low-hanging fruit being picked. In this perspective, coming up with an innovation is like growing fruit, and exploiting an innovation is like eating the fruit over time.

Another school of thought, though, holds that the true work of innovation is not coming up with something big and new, but instead recombining things that already exist. And the more closely we look at how major steps forward in our knowledge and ability to accomplish things have actually occurred, the more this recombinant view makes sense. For example, it's exactly how at least one Nobel Prize–winning innovation came about.

Kary Mullis won the 1993 Nobel Prize in Chemistry for the development of the polymerase chain reaction (PCR), a now ubiquitous technique for replicating DNA sequences. When the idea first came to him on a nighttime drive in California, though, he almost dismissed it out of hand. As he recounted in his Nobel Award speech, "Somehow, I thought, it had to be an illusion. . . . It was too easy. . . . There was not a single unknown in the scheme. Every step involved had been done already."[13] "All" Mullis did was recombine well-understood techniques in biochemistry to generate a new one. And yet it's obvious Mullis's recombination is an enormously valuable one.

After examining many examples of invention, innovation, and technological progress, complexity scholar Brian Arthur became convinced that stories like the invention of PCR are the rule, not the exception. As he summarizes in his book *The Nature of Technology*, "To invent something is to find it in what previously exists."[14] Economist Paul Romer has argued forcefully in favor of this view, the so-called 'new growth theory' within economics, in order to distinguish it from perspectives like Gordon's. Romer's inherently optimistic theory stresses the importance of recombinant innovation. As he writes:

> Economic growth occurs whenever people take resources and rearrange them in ways that make them more valuable. . . . Every generation has perceived the limits to growth that finite resources and undesirable side effects would pose if no new . . . ideas were discovered. And every generation has underestimated the potential for finding new . . . ideas. We consistently fail to grasp how many ideas remain to be discovered. . . . Possibilities do not merely add up; they multiply.[15]

Romer also makes a vital point about a particularly important category of idea, which he calls "meta-ideas":

> Perhaps the most important ideas of all are meta-ideas—ideas about how to support the production and transmission of other ideas. . . . There are . . . two safe predictions. First, the country that takes the lead in the twenty-first century will be the one that implements an innovation that more effectively supports the production of new ideas in the private sector. Second, new meta-ideas of this kind will be found.[16]

Digital Technologies: The Most General Purpose of All

Gordon and Cowen are world-class economists, but they're not giving digital technologies their due. The next great meta-idea, invoked by Romer, has already been found: it can be seen in the new communities of minds and machines made possible by networked digital devices running an astonishing variety of software. The GPT of ICT has given birth to radically new ways to combine and recombine ideas. Like language, printing, the library, or universal education, the global digital network fosters recombinant innovation. We can mix and remix ideas, both old and recent, in ways we never could before. Let's look at a few examples.

Google's Chauffeur project gives new life to an earlier GPT: the internal combustion engine. When an everyday car is equipped with a fast computer and a bunch of sensors (all of which get cheaper according to Moore's Law) and a huge amount of map and street information (available thanks to the digitization of everything) it becomes an autopiloted vehicle straight out of science fiction. While we humans are still the ones doing the driving, innovations like Waze will help us get around more quickly and ease traffic jams. Waze is a recombination of a location sensor, data transmission device (that is, a phone), GPS system, and social network. The team at Waze invented none of these technologies; they just put them together in a new way. Moore's Law made all involved devices cheap enough, and digitization made all necessary data available to facilitate the Waze system.

The Web itself is a pretty straightforward combination of the Internet's much older TCP/IP data transmission network; a markup language called HTML that specified how text, pictures, and so on should be laid out; and a simple PC application called a 'browser' to display the results. None of these elements was particularly novel. Their combination was revolutionary.

Facebook has built on the Web infrastructure by allowing people to digitize their social network and put media online without having to learn HTML. Whether or not this was an intellectually profound combination of technological capabilities, it was a popular and economically significantly one—by July 2013, the company was valued at over $60 billion.[17] When photo sharing became one of the most popular activities on Facebook, Kevin Systrom and Mike Krieger decided to build a smartphone application that mimicked this capability, combining it with the option to modify a photo's appearance with digital filters. This seems like a minor innovation, especially since Facebook already had enabled mobile photo sharing in 2010 when Systrom and Krieger started their project. However, the application they built, called Instagram, attracted more than 30 million users by the spring of 2012, users who had collectively uploaded more than 100 million photographs. Facebook acquired Instagram for approximately $1 billion in April of 2012.

This progression drives home the point that digital innovation is recombinant innovation in its purest form. Each development becomes a building block for future innovations. Progress doesn't run out; it accumulates. And the digital world doesn't respect any boundaries. It extends into the physical one, leading to cars and planes that drive themselves, printers that make parts, and so on. Moore's Law makes computing devices and sensors exponentially cheaper over time, enabling them to be built economically into more and more gear, from doorknobs to greeting cards. Digitization makes available massive bodies of data relevant to almost any situation, and this information can be infinitely reproduced and reused because it is non-rival. As a result of these two forces, the number of potentially valuable building blocks is exploding around the world, and the possibilities are multiplying as never before. We'll call this the 'innovation-as-building-block' view of the world; it's the one held by Arthur, Romer, and the two of us. From this perspective, unlike in the innovation-as-fruit view, building blocks don't ever get eaten or otherwise used up. In fact, they increase the opportunities for future recombinations.

Limits to Recombinant Growth

If this recombinant view of innovation is correct, then a problem looms: as the number of building blocks explodes, the main difficulty is knowing which combinations of them will be valuable. In his paper "Recombinant Growth," the economist Martin Weitzman developed a mathematical model of new growth theory in which the 'fixed factors' in an economy—machine tools, trucks, laboratories, and so on—are augmented over time by pieces of knowledge that he calls 'seed ideas,' and knowledge itself increases over time as previous seed ideas are recombined into new ones.[18] This is an innovation-as-building-block view of the world, where both the knowledge pieces and the seed ideas can be combined and recombined over time.

This model has a fascinating result: because combinatorial possibilities explode so quickly there is soon a virtually infinite number of potentially valuable recombinations of the existing knowledge pieces.* The constraint on the economy's growth then becomes its ability to go through all these potential recombinations to find the truly valuable ones.

As Weitzman writes,

> In such a world the core of economic life could appear increasingly to be centered on the more and more intensive processing of ever-greater numbers of new seed ideas into workable innovations. . . . In the early stages of development, growth is constrained by number of potential new ideas, but later on it is constrained only by the ability to process them.[19]

Gordon asks the provocative question, "Is growth over?" We'll respond on behalf of Weitzman, Romer, and the other new growth theorists with "Not a chance. It's just being held back by our inability to process all the new ideas fast enough."

* Keep in mind that if there are only fifty-two seed ideas in such an economy, they have many more potential combinations than there are atoms in our solar system.

What This Problem Needs Are More Eyeballs and Bigger Computers

If this response is at least somewhat accurate—if it captures something about how innovation and economic growth work in the real world—then the best way to accelerate progress is to increase our capacity to test out new combinations of ideas. One excellent way to do this is to involve more people in this testing process, and digital technologies are making it possible for ever more people to participate. We're interlinked by global ICT, and we have affordable access to masses of data and vast computing power. Today's digital environment, in short, is a playground for large-scale recombination. The open source software advocate Eric Raymond has an optimistic observation: "Given enough eyeballs, all bugs are shallow."[20] The innovation equivalent to this might be, "With more eyeballs, more powerful combinations will be found."

NASA experienced this effect as it was trying to improve its ability to forecast solar flares, or eruptions on the sun's surface. Accuracy and plenty of advance warning are both important here, since solar particle events (or SPEs, as flares are properly known) can bring harmful levels of radiation to unshielded gear and people in space. Despite thirty-five years of research and data on SPEs, however, NASA acknowledged that it had "no method available to predict the onset, intensity or duration of a solar particle event."[21]

The agency eventually posted its data and a description of the challenge of predicting SPEs on Innocentive, an online clearinghouse for scientific problems. Innocentive is 'non-credentialist'; people don't have to be PhDs or work in labs in order to browse the problems, download data, or upload a solution. Anyone can work on problems from any discipline; physicists, for example, are not excluded from digging in on biology problems.

As it turned out, the person with the insight and expertise needed to improve SPE prediction was not part of any recognizable astrophysics community. He was Bruce Cragin, a retired radio frequency engineer living in a small town in New Hampshire. Cragin said that, "Though I hadn't worked in the area of solar physics as such, I had thought a lot about the theory of magnetic reconnection."[22] This was evidently the right theory for the job, because Cragin's approach enabled prediction of SPEs eight hours in advance with 85 percent accuracy, and twenty-four hours in advance with 75 percent accuracy. His recombination of theory and data earned him a thirty-thousand-dollar reward from the space agency.

In recent years, many organizations have adopted NASA's strategy of using technology to open up their innovation challenges and opportunities to more eyeballs. This phenomenon goes by several names, including 'open innovation' and 'crowdsourcing,' and it can be remarkably effective. The innovation scholars Lars Bo Jeppesen and Karim Lakhani studied 166 scientific problems posted to Innocentive, all of which had stumped their home organizations. They found that the crowd assembled around Innocentive was able to solve forty-nine of them, for a success rate of nearly 30 percent. They also found that people whose expertise was far away from the apparent domain of the problem were more likely to submit winning solutions. In other words, it seemed to actually help a solver to be 'marginal'—to have education, training, and experience that were not obviously relevant for the problem. Jeppesen and Lakhani provide vivid examples of this:

> [There were] different winning solutions to the same scientific challenge of identifying a food-grade polymer delivery system by an aerospace physicist, a small agribusiness owner, a transdermal drug delivery specialist, and an industrial scientist. . . . All four submissions successfully achieved the required challenge objectives with differing scientific mechanisms. . . .
>
> [Another case involved] an R&D lab that, even after con-

> sulting with internal and external specialists, did not understand the toxicological significance of a particular pathology that had been observed in an ongoing research program. . . . It was eventually solved, using methods common in her field, by a scientist with a Ph.D. in protein crystallography who would not normally be exposed to toxicology problems or solve such problems on a routine basis.[23]

Like Innocentive, the online startup Kaggle also assembles a diverse, non-credentialist group of people from around the world to work on tough problems submitted by organizations. Instead of scientific challenges, Kaggle specializes in data-intensive ones where the goal is to arrive at a better prediction than the submitting organization's starting baseline prediction. Here again, the results are striking in a couple of ways. For one thing, improvements over the baseline are usually substantial. In one case, Allstate submitted a dataset of vehicle characteristics and asked the Kaggle community to predict which of them would have later personal liability claims filed against them.[24] The contest lasted approximately three months and drew in more than one hundred contestants. The winning prediction was more than 270 percent better than the insurance company's baseline.

Another interesting fact is that the majority of Kaggle contests are won by people who are marginal to the domain of the challenge—who, for example, made the best prediction about hospital readmission rates despite having no experience in health care—and so would not have been consulted as part of any traditional search for solutions. In many cases, these demonstrably capable and successful data scientists acquired their expertise in new and decidedly digital ways.

Between February and September of 2012 Kaggle hosted two competitions about computer grading of student essays, which were sponsored by the Hewlett Foundation.* Kaggle and Hewlett worked

* Improvements in this area are important because essays are better at capturing student leaning than multiple-choice questions, but much more

with multiple education experts to set up the competitions, and as they were preparing to launch many of these people were worried. The first contest was to consist of two rounds. Eleven established educational testing companies would compete against one another in the first round, with members of Kaggle's community of data scientists invited to join in, individually or in teams, in the second. The experts were worried that the Kaggle crowd would simply not be competitive in the second round. After all, each of the testing companies had been working on automatic grading for some time and had devoted substantial resources to the problem. Their hundreds of person-years of accumulated experience and expertise seemed like an insurmountable advantage over a bunch of novices.

They needn't have worried. Many of the 'novices' drawn to the challenge outperformed all of the testing companies in the essay competition. The surprises continued when Kaggle investigated who the top performers were. In both competitions, none of the top three finishers had any previous significant experience with either essay grading or natural language processing. And in the second competition, none of the top three finishers had any formal training in artificial intelligence beyond a free online course offered by Stanford AI faculty and open to anyone in the world who wanted to take it. People all over the world did, and evidently they learned a lot. The top three individual finishers were from, respectively, the United States, Slovenia, and Singapore.

Quirky, another Web-based startup, enlists people to participate in both phases of Weitzman's recombinant innovation—first generating new ideas, then filtering them. It does this by harnessing the power of many eyeballs not only to come up with innovations but also to filter them and get them ready for market. Quirky seeks ideas for new consumer products from its crowd, and also relies on the

---

expensive to grade when human raters are used. Automatic grading of essays would both improve the quality of education and lower its cost.

crowd to vote on submissions, conduct research, suggest improvements, figure out how to name and brand the products, and drive sales. Quirky itself makes the final decisions about which products to launch and handles engineering, manufacturing, and distribution. It keeps 70 percent of all revenue made through its website and distributes the remaining 30 percent to all crowd members involved in the development effort; of this 30 percent, the person submitting the original idea gets 42 percent, those who help with pricing share 10 percent, those who contribute to naming share 5 percent, and so on. By the fall of 2012, Quirky had raised over $90 million in venture capital financing and had agreements to sell its products at several major retailers, including Target and Bed Bath & Beyond. One of its most successful products, a flexible electrical power strip called Pivot Power, sold more than 373 thousand units in less than two years and earned the crowd responsible for its development over $400,000.

Affinnova, yet another young company supporting recombinant innovation, helps its customers with the second of Weitzman's two phases: sorting through the possible combinations of building blocks to find the most valuable ones. It does this by combining crowdsourcing with Nobel Prize–worthy algorithms. When Carlsberg breweries wanted to update the bottle and label for Belgium's Grimbergen, the world's oldest continually produced abbey beer, it knew it had to proceed carefully. The company wanted to update the brand without sacrificing its strong reputation or downplaying its nine hundred years of history. It knew that the redesign would mean generating many candidates for each of several attributes—bottle shape, embossments, label color, label placement, cap design, and so on—then settling on the right combination of all of these. The 'right' combination from among the thousands of possibilities, however, was not obvious at the outset.

The standard approach to this kind of problem is for the design team to generate a few combinations that they think are good, then use focus groups or other small-scale methods to finalize which is best. Affinnova offers a very different approach. It makes use of the mathematics of choice modeling, an advance significant enough to have earned a Nobel Prize for its intellectual godfather, economist Daniel McFadden. Choice modeling quickly identifies people's preferences—do they prefer a brown embossed bottle with a small label, or a green non-embossed one with a large label?—by repeatedly presenting them with a small set of options and asking them to select which they like best. Affinnova presents these options via the Web and can find the mathematically optimal set of options (or at least come close to it) after involving only a few hundred people in the evaluation process. For Grimbergen, the design that resulted from this explicitly recombinant process had an approval rating 3.5 times greater than that of the previous bottle.[25]

When we adopt the perspective of the new growth theorists and match it against what we see with Waze, Innocentive, Kaggle, Quirky, Affinnova, and many others, we become optimistic about the current and future of innovation. And these digital developments are not confined to the high-tech sector—they're not just making computers and networks better and faster. They're helping us drive our cars better (and may soon make it unnecessary for us to drive at all), allowing us to arrive at better predictions of solar flares, solving problems in food science and toxicology, and giving us better power strips and beer bottles. These and countless other innovations will add up over time, and they'll keep coming and keep adding up. Unlike some of our colleagues, we are confident that innovation and productivity will continue to grow at healthy rates in the future. Plenty of building blocks are in place, and they're being recombined in better and better ways all the time.

CHAPTER 6

ARTIFICIAL AND HUMAN INTELLIGENCE IN THE SECOND MACHINE AGE

"And here I am thinking of those astonishing electronic machines . . . by which our mental capacity to calculate and combine is reinforced and multiplied by the process and to a degree that herald . . . astonishing advances."

—Pierre Teilhard de Chardin

THE PREVIOUS FIVE CHAPTERS laid out the outstanding features of the second machine age: sustained exponential improvement in most aspects of computing, extraordinarily large amounts of digitized information, and recombinant innovation. These three forces are yielding breakthroughs that convert science fiction into everyday reality, outstripping even our recent expectations and theories. What's more, there's no end in sight.

The advances we've seen in the past few years, and in the early sections of this book—cars that drive themselves, useful humanoid robots, speech recognition and synthesis systems, 3D printers, *Jeopardy!*-champion computers—are not the crowning achievements of the computer era. They're the warm-up acts. As we move deeper into the second machine age we'll see more and more such wonders, and they'll become more and more impressive.

How can we be so sure? Because the exponential, digital, and recombinant powers of the second machine age have made it possible for humanity to create two of the most important one-time events in our history: the emergence of real, useful artificial intelligence (AI) and the connection of most of the people on the planet via a common digital network.

Either of these advances alone would fundamentally change our growth prospects. When combined, they're more important than anything since the Industrial Revolution, which forever transformed how physical work was done.

Thinking Machines, Available Now

Machines that can complete cognitive tasks are even more important than machines that can accomplish physical ones. And thanks to modern AI we now have them. Our digital machines have escaped their narrow confines and started to demonstrate broad abilities in pattern recognition, complex communication, and other domains that used to be exclusively human.

We've also recently seen great progress in natural language processing, machine learning (the ability of a computer to automatically refine its methods and improve its results as it gets more data), computer vision, simultaneous localization and mapping, and many of the other fundamental challenges of the discipline.

We're going to see artificial intelligence do more and more, and as this happens costs will go down, outcomes will improve, and our lives will get better. Soon countless pieces of AI will be working on our behalf, often in the background. They'll help us in areas ranging from trivial to substantive to life changing. Trivial uses of AI include recognizing our friends' faces in photos and recommending products. More substantive ones include automatically driving cars on the road, guiding robots in warehouses, and better matching jobs and job seekers. But these remarkable advances pale against the life-changing potential of artificial intelligence.

To take just one recent example, innovators at the Israeli company OrCam have combined a small but powerful computer, digital sensors, and excellent algorithms to give key aspects of sight to the visually impaired (a population numbering more than twenty million in the United States alone). A user of the OrCam system, which was introduced in 2013, clips onto her glasses a combination of a tiny digital camera and speaker that works by conducting sound waves through the bones of the head.[1] If she points her finger at a source of text such as a billboard, package of food, or newspaper article, the computer immediately analyzes the images the camera sends to it, then reads the text to her via the speaker.

Reading text 'in the wild'—in a variety of fonts, sizes, surfaces, and lighting conditions—has historically been yet another area where humans outpaced even the most advanced hardware and software. OrCam and similar innovations show that this is no longer the case, and that here again technology is racing ahead. As it does, it will help millions of people lead fuller lives. The OrCam costs about $2,500—the price of a good hearing aid—and is certain to become cheaper over time.

Digital technologies are also restoring hearing to the deaf via cochlear implants and will probably bring sight back to the fully blind; the FDA recently approved a first-generation retinal implant.[2] AI's benefits extend even to quadriplegics, since wheelchairs can now be controlled by thoughts.[3] Considered objectively, these advances are something close to miracles—and they're still in their infancy.

Artificial intelligence will not just improve lives; it will also save them. After winning *Jeopardy!*, for example, Watson enrolled in medical school. To be a bit more precise, IBM is applying the same innovations that allowed Watson to answer tough questions correctly to the task of helping doctors better diagnose what's wrong with their patients. Instead of volumes and volumes of general knowledge, the supercomputer is being trained to sit on top of all of the world's high-quality published medical information; match it against patients' symptoms, medical histories, and test results; and formulate both a diagnosis and a treatment plan. The huge amounts of information involved in modern medicine make this type of advance critically important. IBM estimates that it would take a human doctor 160 hours of reading each and every week just to keep up with relevant new literature.[4]

IBM and partners including Memorial Sloan-Kettering Cancer Center and the Cleveland Clinic are working to build Dr. Watson. The organizations involved in this program are careful to stress that the AI technologies will be used to augment physicians' clinical expertise and judgment, not replace them. Still, it is not implausible that Dr. Watson might one day be the world's best diagnostician.

We're already seeing AI-aided diagnoses in some medical specialties. A team led by pathologist Andrew Beck developed the C-Path (computational pathologist) system to automatically diagnose breast cancer and predict survival rates by examining images of tissue, just as human pathologists do.[5] Since the 1920s, these humans have been trained to look at the same small set of cancer cell features.[6] The C-Path team, in contrast, had its software look at images with a fresh eye—without any pre-programmed notions about which features were associated with cancer severity or patient prognosis. Not only was this software at least as accurate as humans, it also identified three features of breast cancer tissue that turned out to be good predictors of survival rates. Pathologists, however, had not been trained to look for them.

As it races ahead, artificial intelligence might bring with it some troubles, which we'll discuss in our conclusion. But fundamentally, the development of thinking machines is an incredibly positive one.

Billions of Innovators, Coming Soon

In addition to powerful and useful AI, the other recent development that promises to further accelerate the second machine age is the digital interconnection of the planet's people. There is no better resource for improving the world and bettering the state of humanity than the world's humans—all 7.1 billion of us. Our good ideas and innovations will address the challenges that arise, improve the quality of our lives, allow us to live more lightly on the planet, and help us take better care of one another. It is a remarkable and unmistakable fact that, with the exception of climate change, virtually all environmental, social, and individual indicators of health have improved over time, even as human population has increased.

This improvement is not a lucky coincidence; it is cause and effect. Things have gotten better *because* there are more people, who in total have more good ideas that improve our overall lot. The economist Julian Simon was one of the first to make this optimistic argument, and he advanced it repeatedly and forcefully throughout his career. He wrote, "It is your mind that matters economically, as much or more than your mouth or hands. In the long run, the most important economic effect of population size and growth is the contribution of additional people to our stock of useful knowledge. And this contribution is large enough in the long run to overcome all the costs of population growth."[7]

Both theory and data bear out Simon's insight. The theory of recombinant innovation stresses how important it is to have more eyeballs looking at challenges and more brains thinking about how existing building blocks can be rearranged to meet them. This theory further holds that people also play the vital role of filtering and improving the innovations of others. And the data on everything from air quality to commodity prices to levels of violence show improvement over time. These data, in other words, show humanity's remarkable ability to meet its challenges.

We do have one quibble with Simon, however. He wrote that, "The main fuel to speed the world's progress is our stock of knowledge, and the brake is our lack of imagination."[8] We agree about the fuel but disagree about the brake. The main impediment to progress has been that, until quite recently, a sizable portion of the world's people had no effective way to access the world's stock of knowledge or to add to it.

In the industrialized West we have long been accustomed to having libraries, telephones, and computers at our disposal, but these have been unimaginable luxuries to the people of the developing world. That situation is rapidly changing. In 2000, for example, there were approximately seven hundred million mobile phone subscriptions in the world, fewer than 30 percent of which were in developing countries.[9] By 2012 there were more than six billion subscriptions, over 75 percent of which were in the developing world. The World Bank estimates that three-quarters of the people on the planet now have access to a mobile phone, and that in some countries mobile telephony is more widespread than electricity or clean water.

The first mobile phones bought and sold in the developing world were capable of little more than voice calls and text messages, yet even these simple devices could make a significant difference. Between 1997 and 2001 the economist Robert Jensen studied a set of coastal villages in Kerala, India, where fishing was the main industry.[10] Jensen gathered data both before and after mobile phone service was introduced, and the changes he documented are remarkable. Fish prices stabilized immediately after phones were introduced, and even though these prices dropped on average, fishermen's profits actually increased because they were able to eliminate the waste that occurred when they took their fish to markets that already had enough supply for the day. The overall economic well-being of both buyers and sellers improved, and Jensen was able to tie these gains directly to the phones themselves.

Now, of course, even the most basic phones sold in the developing world are more powerful than the ones used by Kerala's fisherman over a decade ago. Approximately 70 percent of all phones sold worldwide in 2012 were 'feature phones'—less capable than the Apple iPhone and Samsung Galaxy smartphones of the rich world, but still able to take pictures (and often videos), browse the Web, and run at least some applications.[11] And cheap mobile devices keep improving. Technology analysis firm IDC forecasts that smartphones will outsell feature phones in the near future, and will make up about two-thirds of all sales by 2017.[12]

This shift is due to continued simultaneous performance improvements and cost declines in both mobile phone devices and networks, and it has an important consequence: it will bring billions of people into the community of potential knowledge creators, problem solvers, and innovators.

Today, people with connected smartphones or tablets anywhere in the world have access to many (if not most) of the same communication resources and information that we do while sitting in our offices at MIT. They can search the Web and browse Wikipedia. They can follow online courses, some of them taught by the best in the academic world. They can share their insights on blogs, Facebook, Twitter, and many other services, most of which are free. They can even conduct sophisticated data analyses using cloud resources such as Amazon Web Services and R, an open source application for statistics.[13] In short, they can be full contributors in the work of innovation and knowledge creation, taking advantage of what Autodesk CEO Carl Bass calls "infinite computing."[14]

Until quite recently rapid communication, information acquisition, and knowledge sharing, especially over long distances, were essentially limited to the planet's elite. Now they're much more democratic and egalitarian, and getting more so all the time. The journalist A. J. Liebling famously remarked that, "Freedom of the press is limited to those who own one." It is no exaggeration to say that billions of people will soon have a printing press, reference library, school, and computer all at their fingertips.[15]

Those of us who believe in the power of recombinant innovation believe that this development will boost human progress. We can't predict exactly what new insights, products, and solutions will arrive in the coming years, but we are fully confident that they'll be impressive. The second machine age will be characterized by countless instances of machine intelligence and billions of interconnected brains working together to better understand and improve our world. It will make mockery out of all that came before.

CHAPTER 7

COMPUTING BOUNTY

"Most economic fallacies derive from the tendency to assume that there is a fixed pie, that one party can gain only at the expense of another."

—Milton Friedman

EACH DAY GOVERNMENT AGENCIES, think tanks, NGOs, and academic researchers generate more statistics than any person could read, let alone absorb. On television, in the pages of the business press, and in the blogosphere, a chorus of analysts debate and predict trends in interest rates, unemployment, stock prices, deficits and myriad other indicators. But when you zoom out and consider trends over the past century, one overwhelming fact looms above all others: overall living standards have increased enormously in the United States and worldwide. In the United States, the rate of GDP growth per person has averaged 1.9 percent per year going back to the early 1800s.[1] Applying the rule of 70 (the time to double a value is roughly equal to 70 divided by its growth rate), we see that this was enough to double living standards every thirty-six years, quadrupling them over the course of a typical lifetime.*

This increase is important because economic growth can help solve a host of other challenges. If GDP of the United States grows just 1 percent faster each year than currently projected, Americans would be five trillion dollars richer by 2033.[2] If GDP grows just 0.5 percent faster, the U.S. budget problem would be solved without any

* The Rule of 70 (or, more precisely, the rule of 69.3 percent) is based on the following equation: $(1 + x)^y = 2$ where x is the rate of growth and y is the number of years. Taking the natural logarithm of both sides gives $y \ln(1 + x) = \ln 2$. The $\ln(2)$ is 0.693 and for small x, $\ln(1 + x)$ is roughly equal to x, so the equation simplifies to $xy = 70$ percent.

changes to policy.[3] Of course, slower growth would make it significantly harder to close the deficit, let alone increase spending on any new initiatives or cut taxes.

Productivity Growth

But what drives increases in GDP per person? Part of it comes from using more resources. But most of it comes from increases in our ability to get more output from the given level of inputs—in other words, increases in productivity. (Most commonly, this term is used as shorthand for 'labor productivity,' which is output per hour worked [or output per worker].) * In turn, productivity growth comes from innovations in technology and techniques of production.

Simply working more hours does not increase productivity. Indeed, Americans once routinely worked fifty, sixty, or even seventy hours per week. While some still do, the average workweek is shorter now (thirty-five hours per week), and yet living standards are higher. Robert Solow got his Nobel Prize in Economics for showing that increases in labor input and capital input could not explain most of the increase in the total output of the economy.† In fact, it would take the average American only eleven hours of labor per week to produce as much as he or she produced in forty hours in 1950. That

* One can also measure capital productivity, which is output per unit of capital input; or multifactor productivity, which is output divided by a weighted average of both capital and labor inputs. Economists sometimes use another term for multifactor productivity, the "Solow Residual," which better reflects the fact that we don't necessarily know its origins. Robert Solow himself noted that it was less a concrete measure of technological progress than a "measure of our ignorance."

† That's a good thing, because there are natural limits to how much we can increase inputs, especially labor. They're subject to diminishing returns—no one is going to work more than twenty-four hours a day, or employ more than 100 percent of the labor force. In contrast, productivity growth reflects ability to innovate—it's limited only by our imaginations.

FIGURE 7.1 Labor Productivity

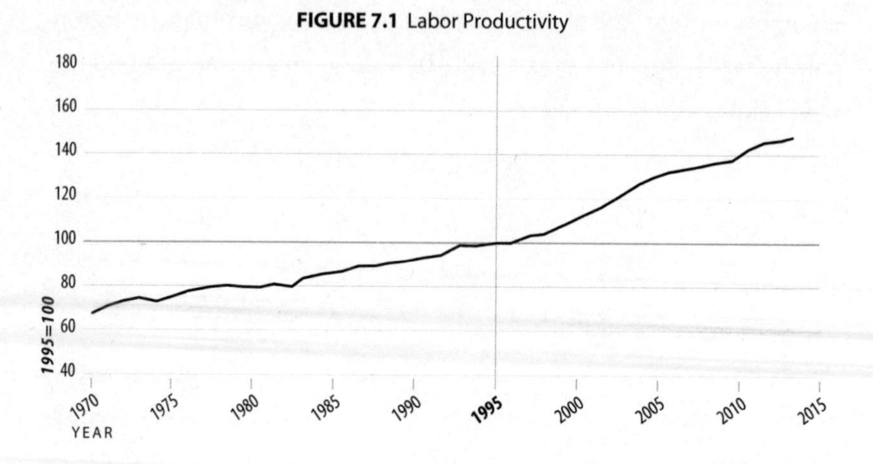

rate of improvement is comparable for workers in Europe and Japan, and even higher in some developing nations.*

Productivity improvement was particularly rapid in the middle part of the twentieth century, especially the 1940s, 50s, and 60s, as the technologies of the first machine age, from electricity to the internal combustion engine, started firing on all cylinders. However, in 1973 productivity growth slowed down (see figure 7.1).

In 1987, Bob Solow himself noted that the slowdown seemed to coincide with the early days of the computer revolution, famously remarking, "We see the computer age everywhere, except in the productivity statistics."[4] In 1993, Erik published an article evaluating the "Productivity Paradox" that noted the computers were still a small share of the economy and that complementary innovations were typically needed before general purpose technologies like IT had their real impact.[5] Later work taking into account more detailed

* Output divided by labor and physical capital inputs is often more ambitiously called 'total factor productivity.' However, that term can be a bit misleading, because there are other inputs to production. For instance, companies can make major investments in intangible organizational capital. The more kinds of inputs we are able to measure, the better we can account for overall output growth. As a result, the residual that we label "productivity" (not explained by growth of inputs) will get smaller.

data on productivity and IT use among individual firms revealed a strong and significant correlation: the heaviest IT users were dramatically more productive than their competitors.[6] By the mid-1990s, these benefits were big enough to become visible in the overall U.S. economy, which experienced a general productivity surge. While this rise had a number of causes, economists now attribute the lion's share of those gains to the power of IT.[7]

The productivity slowdown in the 1970s, and the subsequent speed-up twenty years later, had an interesting precedent. In the late 1890s, electricity was being introduced to American factories. But the "productivity paradox" of that era was that labor productivity growth did not take off for over twenty years. While the technologies involved were very different, many of the underlying dynamics were quite similar.

University of Chicago economist Chad Syverson looked closely at the underlying productivity data and showed how eerily close this analogy is.[8] As shown in figure 7.2, the slow start and subsequent acceleration of productivity growth in the electricity era matches well with the speed-up that began in the 1990s. The key to understanding this pattern is the realization that, as discussed in chapter

FIGURE 7.2 Labor Productivity in Two Eras

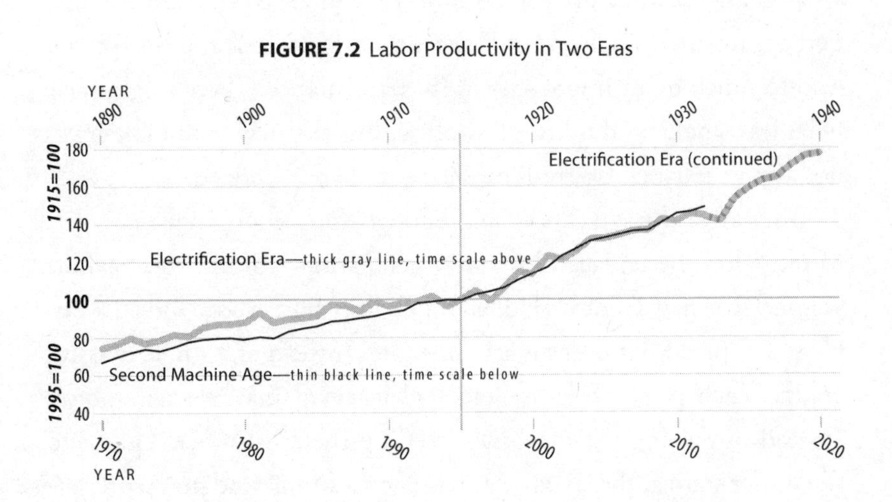

5, GPTs always need complements. Coming up with those can take years, or even decades, and this creates lags between the introduction of a technology and the productivity benefits. We've clearly seen this with both electrification and computerization.

Perhaps the most important complementary innovations are the business process changes and organizational coinventions that new technologies make possible. Paul David, an economic historian at Stanford University and the University of Oxford, examined the records of American factories when they first electrified and found that they often retained a similar layout and organization to those that were powered by steam engines.[9] In a steam engine–driven plant, power was transmitted via a large central axle, which in turn drove a series of pulleys, gears, and smaller crankshafts. If the axle was too long the torsion involved would break it, so machines needed to be clustered near the main power source, with those requiring the most power positioned closest. Exploiting all three dimensions, industrial engineers put equipment on floors above and below the central steam engines to minimize the distances involved.

Years later, when that hallowed GPT electricity replaced the steam engine, engineers simply bought the largest electric motors they could find and stuck them where the steam engines used to be. Even when brand-new factories were built, they followed the same design. Perhaps unsurprisingly, records show that the electric motors did not lead to much of an improvement in performance. There might have been less smoke and a little less noise, but the new technology was not always reliable. Overall, productivity barely budged.

Only after thirty years—long enough for the original managers to retire and be replaced by a new generation—did factory layouts change. The new factories looked much like those we see today: a single story spread out over an acre or more. Instead of a single massive engine, each piece of equipment had its own small electric motor. Instead of putting the machines needing the most power closest to the power source, the layout was based on a simple and powerful new principle: the natural workflow of materials.

Productivity didn't merely inch upward on the resulting assembly lines; it doubled or even tripled. What's more, for most of the subsequent century, additional complementary innovations, from lean manufacturing and steel minimills to Total Quality Management and Six Sigma principles, continued to boost manufacturing productivity.

As with earlier GPTs, significant organizational innovation is required to capture the full benefit of second machine age technologies. Tim Berners-Lee's invention of the World Wide Web in 1989, to take an obvious example, initially benefited only a small group of particle physicists. But due in part to the power of digitization and networks to speed the diffusion of ideas, complementary innovations are happening faster than they did in the first machine age. Less than ten years after its introduction, entrepreneurs were finding ways to use the Web to reinvent publishing and retailing.

While less visible, the large enterprise-wide IT systems that companies rolled out in the 1990s have had an even bigger impact on productivity.[10] They did this mainly by making possible a wave of business process redesign. For example, Walmart drove remarkable efficiencies in retailing by introducing systems that shared point-of-sale data with their suppliers. The real key was the introduction of complementary process innovations like vendor managed inventory, cross-docking, and efficient consumer response that have become staple business-school case studies. They not only made it possible to increase sales from $1 billion a week in 1993 to $1 billion every thirty-six hours in 2001, but also helped drive dramatic increases in the entire retailing and distribution industries, accounting for much of the additional productivity growth nationwide during this period.[11]

IT investment soared in the 1990s, peaking with a surge of investment in the latter half of the decade as many companies upgraded their systems to take advantage of the Internet, implement large enterprise systems, and avoid the much-hyped Y2K bug. At the same time, innovation in semiconductors took gigantic leaps, so the surging spending on IT delivered even more rapidly increasing levels of computer power. A decade after the computer productivity paradox was popularized, Harvard's Dale Jorgenson, working with Kevin Stiroh at the New York Federal Reserve Bank did a careful growth accounting and concluded, "A consensus has emerged that a large portion of the acceleration through 2000 can be traced to the sectors of the economy that produce information technology or use IT equipment and software most intensively."[12] But it's not just the computer-producing sectors that are doing well. Kevin Stiroh of the New York Federal Reserve Bank found that industries that were heavier *users* of IT tended to be more productive throughout the 1990s. This pattern was even more evident in recent years, according to a careful study by Harvard's Dale Jorgenson and two coauthors. They found that total factor productivity growth increased more between the 1990s and 2000s in IT-using industries, while it fell slightly in those sectors of the economy that did not use IT extensively.[13]

It's important to note that the correlation between computers and productivity is not just evident at the industry level; it occurs at the level of individual firms as well. In work Erik did with Lorin Hitt of the University of Pennsylvania Wharton School, he found that firms that use more IT tend to have higher levels of productivity and faster productivity growth than their industry competitors.[14]

The first five years of the twenty-first century saw a renewed wave of innovation and investment, this time less focused on computer hardware and more focused on a diversified set of applications and process innovations. For instance, as Andy described in a case study he did for Harvard Business School, CVS found that their prescription drug ordering process was a source of customer frustration, so they redesigned and simplified it.[15] By embedding the steps in an enterprise-wide software system, they were able to replicate the drug ordering process in over four thousand locations, dramatically boosting customer satisfaction and ultimately profits. CVS was not atypical. In a statistical analysis of over six hundred firms that Erik did with Lorin Hitt, he found it takes an average five to seven years before full productivity benefits of computers are visible in the productivity of the firms making the investments. This reflects the time and effort required to make the other complementary investments that bring a computerization effort success. In fact, for every dollar of investment in computer hardware, companies need to invest up to another nine dollars in software, training, and business process redesign.[16]

The effects of organizational changes like these became increasingly visible in the industry-level productivity statistics.[17] The productivity surge in the 1990s was most visible in computer-producing industries, but overall productivity grew even faster in the early years of the twenty-first century, when a much broader set of industries saw significant productivity gains. Like earlier GPTs, the power of computers was their ability to affect productivity far from their 'home' industry.

Overall, American productivity growth in the decade following the year 2000 exceeded even the high growth rates of the roaring 1990s, which in turn was higher than 1970s or 1980s growth rates had been.[18]

Today American workers are more productive than they've ever been, but a closer look at recent numbers tells a more nuanced story. The good performance since the year 2000 was clustered in the early years of the decade. Since 2005, productivity growth has not been as strong. As noted in chapter 5, this has led to a new wave of worries about the "end of growth" by economists, journalists, and bloggers. We are not convinced by the pessimists. The productivity lull after the introduction of electricity did not mean the end of growth, nor did the lull in the 1970s.

Part of the recent slowdown simply reflects the Great Recession and its aftermath. Recessions are always times of pessimism, which is understandable, and the pessimism invariably spills over into predictions about technology and the future. The financial crisis and burst of the housing bubble led to a collapse of consumer confidence and wealth, which translated into dramatically lower demand and GDP. While the recession technically ended in June 2009, as we write this in 2013 the U.S. economy is still operating well below its potential, with unemployment at 7.6 percent and capacity utilization at 78 percent. During such a slump, any metric that includes output in the numerator, such as labor productivity, will often be at least temporarily depressed. In fact, when you look at history, you see that in the early years of the Great Depression, in the 1930s, productivity didn't just slow but actually fell for two years in a row—something it never did in the recent slump. Growth pessimists had even more company in the 1930s than they do today, but the following three decades proved to be the best ones of the twentieth century. Go back to figure 7.2 and look most closely at the dashed line charting the years following the dip in productivity in the early 1930s. You'll see the biggest wave of growth and bounty that the first machine age ever delivered.

The explanation for this productivity surge is in the lags that we always see when GPTs are installed. The benefits of electrification stretched for nearly a century as more and more complementary innovations were implemented. The digital GPTs of the second machine age are no less profound. Even if Moore's Law ground to a halt today, we could expect decades of complementary innovations to unfold and continue to boost productivity. However, unlike the steam engine or electricity, second machine age technologies continue to improve at a remarkably rapid exponential pace, replicating their power with digital perfection and creating even more opportunities for combinatorial innovation. The path won't be smooth—for one thing, we haven't banished the business cycle—but the fundamentals are in place for bounty that vastly exceeds anything we've ever seen before.

CHAPTER 8

BEYOND GDP

"The Gross National Product does not include the beauty of our poetry or the intelligence of our public debate. It measures neither our wit nor our courage, neither our wisdom nor our learning, neither our compassion nor our devotion. It measures everything, in short, except that which makes life worthwhile."

—Robert F. Kennedy

WHEN PRESIDENT HOOVER WAS trying to understand what was happening during the Great Depression and design a program to fight it, a comprehensive system of national accounts did not exist. He had to rely on scattered data like freight car loadings, commodity prices, and stock price indexes that gave only an incomplete and often unreliable view of economic activity. The first set of national accounts was presented to Congress in 1937 based on the pioneering work of Nobel Prize winner Simon Kuznets, who worked with researchers at the National Bureau of Economic Research and a team at the U.S. Department of Commerce. The resulting set of metrics have served as beacons that helped illuminate many of the dramatic changes that transformed the economy throughout the twentieth century.

But as the economy has changed so, too, must our metrics. More and more what we care about in the second machine age are ideas, not things—mind, not matter; bits, not atoms; and interactions, not transactions. The great irony of this information age is that, in many ways, we actually know less about the sources of value in the economy than we did fifty years ago. In fact, much of the change has been invisible for a long time simply because we did not know what to look for. There's a huge layer of the economy unseen in the official data and, for that matter, unaccounted for on the income statements and balance sheets of most companies. Free digital goods, the sharing economy, intangibles and changes in our relationships have already had big effects on our well-being. They also call for new organizational structures, new skills, new institutions, and perhaps even a reassessment of some of our values.

Music to Your Ears

The story of music's move from physical media to computer files has been told often and well, but one of that transition's most interesting aspects is less discussed. Music is hiding itself from our traditional economic statistics. Sales of music on physical media declined from 800 million units in 2004 to less than 400 million units in 2008. Yet over the same time period total units of music purchased still grew, reflecting an even faster increase in the purchases of digital downloads. Digital streams such as iTunes, Spotify, or Pandora also came to prominence, and, of course, the purchase data don't reflect the even larger number of songs that were shared, streamed, or downloaded for free, often via piracy. Before the rise of the MP3, even the most fanatical music fan, with a basement stacked high with LPs, tapes, and CDs, wouldn't have had a fraction of the twenty million songs available on a child's smartphone via services like Spotify or Rhapsody. What's more, clever research by Joel Waldfogel at the University of Minnesota finds quantitative evidence that the overall quality of music has not declined over the past decade and is, if anything, higher than ever.[1] If you're like most people, you are listening to more and better music than ever before.

So how did music disappear? The value of music has not changed, only the price. From 2004 to 2008, the combined revenue from sales of music dropped from $12.3 billion to $7.4 billion—that's a decline of 40 percent. Even when we include all digital sales, throwing in ringtones on mobile phones for good measure, the total revenues to the record companies are still down 30 percent.

Similar economics apply when you read the *New York Times*, *Bloomberg Businessweek*, or *MIT Sloan Management Review* online at a reduced price or for free instead of buying a physical copy at the newsstand, or when you use Craigslist instead of the classified ads, or when you share photos via Facebook instead of mailing prints around to friends and relatives. Analog dollars are becoming digital pennies.

By now, the number of pages of digital text and images on the Web is estimated to exceed one trillion.[2] As discussed in chapter 4, bits are created at virtually zero cost and transmitted almost instantaneously worldwide. What's more, a copy of a digital good is exactly identical to the original. This leads to some very different economics and some special measurement problems. When a business traveler calls home to talk to her children via Skype, that may add zero to GDP, but it's hardly worthless. Even the wealthiest robber baron would have been unable to buy this service. How do we measure the benefits of free goods or services that were unavailable at any price in previous eras?

What GDP Leaves Out

Despite all the attention it gets from economists, pundits, journalist, and politicians, GDP, even if it were perfectly measured, does not quantify our welfare. The trends in GDP growth and productivity growth covered in chapter 7 are important, but they are not sufficient measures of our overall well-being, or even our economic well-being. Robert Kennedy put this poetically in his quote at the beginning of this chapter.

While it would be unrealistic to put a dollar value on stirring oratory like RFK's, we can do a better job of understanding our basic economic progress by considering some of the changes in the goods and services that we are able to consume. It soon becomes clear that the trends in the official statistics not only underestimate our bounty, but in the second machine age they have also become increasingly misleading.

In addition to their vast library of music, children with smartphones today have access to more information in real time via the mobile web than the president of the United States had twenty years ago. Wikipedia alone claims to have over fifty times as much information as *Encyclopaedia Britannica*, the premier compilation of knowledge for most of the twentieth century.[3] Like Wikipedia but unlike *Britannica*, much of the information and entertainment available today is free, as are over one million apps on smartphones.[4]

Because they have zero price, these services are virtually invisible in the official statistics. They add value to the economy, but not dollars to GDP. And because our productivity data are, in turn, based on GDP metrics, the burgeoning availability of free goods does not move the productivity dial. There's little doubt, however, that they have real value. When a girl clicks on a YouTube video instead of going to the movies, she's saying that she gets more net value from YouTube than traditional cinema. When her brother downloads a free gaming app on his iPad instead of buying a new video game, he's making a similar statement.

Free: Good for Well-Being, Bad for GDP

In some ways, the proliferation of free products even pushes GDP downward. If the cost of creating and delivering an encyclopedia to your desktop is a few pennies instead of thousands of dollars, then you're certainly better off. But this decrease in costs *lowers* GDP even as our personal well-being increases, leaving GDP to travel in the opposite direction of our true well-being. A simple switch to using a free texting service like Apple's iChat instead of SMS, free classifieds like Craigslist instead of newspaper ads, or free calls like Skype instead of a traditional telephone service can make billions of dollars disappear from companies' revenues and the GDP statistics.[5]

As these examples show, our economic welfare is only loosely related to GDP. Unfortunately many economists, journalists, and much of the general public still use "GDP growth" as a synonym for "economic growth." For much of the twentieth century, this was a fair comparison. If one assumes that each additional unit of production created a similar increment in well-being, then counting up how many units were produced, as GDP does, would be a fine approximation of welfare. A nation that sells more cars, more bushels of wheat, and more tons of steel probably corresponds to a nation whose people are better off.

With a greater volume of digital goods introduced each year that do not have a dollar price, this traditional GDP heuristic is becoming less useful. As we discussed in chapter 4, the second machine age is often described as an "information economy," and with good reason. More people than ever are using Wikipedia, Facebook, Craigslist, Pandora, Hulu, and Google, with thousands of new digital goods introduced each year.

The U.S. Bureau of Economic Analysis defines the information sector's contribution to the economy as the sum of the sales of software, publishing, motion pictures, sound recording, broadcasting, telecommunications, and information and data processing services. According to the official measures, these account for just 4 percent of our GDP today, almost precisely the same share of GDP as in the late 1980s, before the World Wide Web was even invented. But clearly this isn't right. The official statistics are missing a growing share of the real value created in our economy.

Measuring Growth with a Time Machine: Would You Rather . . . ?

Can we improve on GDP as a measure of well-being? Economists sometimes use an alternate approach that resembles the children's game "Would you rather . . . ?" The 1912 Sears shopping catalog had thousands of items for sale, from a "Sears Motor Car" for $335 (page 1,213) to dozens of pairs of women's shoes, some available for as little as $1.50 (pages 371–79). Suppose I gave you an expanded version of this catalog that listed *all* the goods and services available in 1912, not just from Sears, but from any seller in the economy of 1912, and all the same prices as 1912.[6] Would you rather shop exclusively in that old catalog, with no other choices, or would you rather pay today's prices for a full selection of today's goods and services?

Or to make the comparison less difficult, pick two more recent catalogs, like 1993 versus 2013. If you had fifty thousand dollars to spend, would you rather be able to buy any 1993-model car (it would be brand-new) and pay 1993 prices, or a 2013 car and pay 2013 prices? Would you rather be able to buy the bananas, contact lenses, chicken wings, shirts, chairs, banking services, airline tickets, movies, telephone service, health care, housing services, light bulbs, computers, gasoline, and other goods and services that were available in 1993 at 1993 prices? Or would you rather buy the equivalent 2013 basket of services at 2013 prices?

Bananas or a gallon of gasoline have not really changed qualitatively since 1993, so the only difference to consider is their price. If that were the only difference, inflation would be easy to calculate, and the "would you rather" comparison would be a lot easier, too. For other goods, though, especially second machine age goods like online information and mobile phone capabilities, there have been big changes in quality, so the real quality-adjusted price may have fallen even if the nominal sticker price has increased. What's more, there are a lot of new goods that didn't exist before, especially digital goods. There are also some older goods and services that have been discontinued or degraded. It's hard to find a good horsehide razor strop these days,[7] or a 1993 vintage personal computer, or a gas station where the attendants routinely wash your windshield for no charge, like they once did.

Once you pick which catalog you like better, the next step asks how much money I would have to pay you to make you indifferent between the two catalogs. If I have to pay you 20 percent more to make you just as happy shopping from the new catalog as you would be shopping from the old catalog, then the overall price index has increased by 20 percent. And if your income has not changed, then that erosion of purchasing power translates to an equivalent fall in your standard of living. Similarly, if your income increases faster than the price index, then your standard of living is increasing.

This approach makes sense conceptually, and it's the basis for the way most modern governments calculate changes in the standard of living. For instance, the cost of living adjustments used to index Social Security payments are based on this kind of analysis.[8] But the data used for these calculations are almost always drawn, understandably, from market transactions where money changes hands. The free economy is not factored in.

Consumer Surplus: How Much Would You Pay If You Had To?

An alternative approach measures the consumer surplus generated by goods and services. Consumer surplus compares the amount a consumer would have been willing to pay for something to the amount they actually have to pay. If you would happily pay one dollar to read the morning newspaper but instead you get it for free, then you've just gained one dollar of consumer surplus. However, as noted above, replacing a paid newspaper with an equivalent free new service would *decrease* GDP even though it *increased* consumer surplus.[9] In this case, consumer surplus would be a better measure of our economic well-being. Yet as appealing as consumer surplus is as a concept, it is also extremely difficult to measure.

The difficulty in measuring the consumer surplus, however, has not stopped a number of researchers from trying to eke out some estimates. In 1993, Erik wrote a paper calculating that the rapidly growing consumer surplus from price declines in computers increased economic welfare by about $50 billion each year.*[10]

* There have been a number of related findings since then. Last year, the economists Jeremy Greenwood and Karen Kopecky applied a similar approach and found a similar growth contribution for personal computers

Of course, when the product being studied is already free, looking at price declines doesn't work. Recent research that Erik did with Joo Hee Oh, a postdoctoral student at MIT, took a different approach. They started with the observation that even when people don't pay with money, they still give up something valuable whenever they use their Internet: their time.[11] No matter how rich or poor we are, each of us gets twenty-four hours in a day. In order to consume YouTube, Facebook, or e-mail, we must 'pay' attention. In fact, Americans nearly doubled the amount of leisure time they spent on Internet between 2000 and 2011. This implies that they valued it more than the other ways they could spend their time. By considering the value of users' time and comparing leisure time spent on the Internet to time spent in other ways, Erik and Joo Hee estimated that the Internet created about $2,600 of value per user each year. None of this showed up in the GDP statistics but if it had, GDP growth—and thus productivity growth—would have been about 0.3 percent higher each year. In other words, instead of the reported 1.2 percent productivity growth for 2012, it would have been 1.5 percent.

In contrast to leisure, where more time is a good thing, value at work is created by saving time. Hal Varian, the chief economist at Google, looked specifically at time savings gained from Google searches.[12] He and his team gathered a random sample of Google queries, such as: "In making cookies, does the use of butter or mar-

---

alone. Shane Greenstein and Ryan McDevitt, another pair of economists, asked how much consumer surplus was created by the spread of broadband Internet access. They looked at how the real price of broadband had declined over time and how adoption of the service had increased. From that, they estimate how much people would have been willing to pay compared to what they actually paid, and thus arrive at the consumer surplus. A research team at McKinsey took a more direct approach. The team asked 3,360 consumers what they would have been willing to pay for sixteen specific services available via the Internet. The average willingness to pay added up to fifty dollars per month. Based on this, the team estimated that Americans received over $35 billion worth of consumer surplus from the free Internet. The biggest single category was e-mail, with social networks like Facebook close behind.

garine affect the size of the cookie?" The team then did their best job to answer the questions without using Google—by looking answers up in the library, for instance. On average it took about twenty-two minutes to answer a query without Google (not counting travel time to the library!) but only seven minutes to answer the same query with Google. Google saved an average of fifteen minutes per query. When you multiply that time difference out across all the queries that the average American makes using the average hourly wage of Americans ($22), that works out to about $500 per adult worker per year.

As anyone who has been caught up in the pleasures of surfing the Web (perhaps while 'doing research' for a book) can attest, though, the strict distinction between work and play or input and output that economists make is not always so clear. The billions of hours that people spend uploading, tagging, and commenting on photos on social media sites like Facebook unquestionably creates value for their friends, family, and even strangers. Yet at the same time these hours are uncompensated, so presumably the people doing this 'work' find it more intrinsically rewarding than the next best use of their time. To get a sense on the scale of this effort, consider that last year users collectively spent about 200 million hours each day just on Facebook, much of it creating content for other users to consume.[13] That's ten times as many person-hours as were needed to build the entire Panama Canal.[14] None of this is counted in our GDP statistics as either input or output, but these kinds of zero-wage and zero-price activities still contribute to welfare. Researchers like Luis von Ahn at Carnegie Mellon are working on ways of motivating and organizing millions of people to create value via collective projects on the Internet.[15]

New Goods and Services

In the early days of the 1990s Internet boom, venture capitalists used to joke that there were only two numbers in the new economy: infinity and zero. Certainly, a big part of the value in the new economy has come from the reduction in the price of many goods to zero. But what about the other end of that spectrum, price drops from infinity down to some finite number? Suppose Warner Bros. makes a new movie and you can watch it for nine dollars. Has your welfare increased? Before the movie was conceived, cast, filmed, and distributed, you couldn't buy it at any price, even infinity. In a sense, paying just nine bucks is a pretty large price reduction from infinity, or whatever the maximum price was that you would have been willing to pay. Similarly, we now have access to all sorts of new services that never existed before, some of which we saw in earlier chapters. Much of the increase in our welfare over the past century comes not just from making existing goods more cheaply but from expanding the range of goods and services available.

Seventy-seven percent of software companies report the introduction of new products each year, and Internet retailing has vastly expanded the set of goods available to most consumers.[16] With a few clicks, over two million books can be found and purchased at Amazon.com. By contrast, the typical physical bookstore has about 40,000 titles and even the largest Barnes & Noble store in New York City stocks only 250,000 titles. As documented in a research paper that Erik wrote with Michael Smith and Jeffrey Hu, there have been similar increases in the online selection for other categories such as videos, music, electronics, and collectibles. Every time a new product is made available, it increases consumer surplus.

One way to think of the value created is to imagine that the new product always existed, but only at such a high price that no one could buy it. Making it available is like lowering the price to a more reasonable level. There have even been substantial increases in the number of stock keeping units (SKUs) in most physical stores as computerized inventory management systems, supply chains, and manufacturing have become more efficient and flexible. For the overall economy, the official GDP numbers miss the value of new goods and services added to the tune of about 0.4 percent of additional growth each year, according to economist Robert Gordon.* Remember that productivity growth has been in the neighborhood of 2 percent per year for most of the past century, so contribution of new goods is not a trivial portion.

Reputations and Recommendations

Digitization also brings a related but subtler benefit to the vast array of goods and services that already exist in the economy. Lower search and transaction costs mean faster and easier access and increased efficiency and convenience. For example, the rating site Yelp collects millions of customer reviews to help diners find nearby restaurants in the quality and price ranges they seek, even when they are visiting new cities. The reservation service OpenTable then lets them book a table with just a few mouse clicks.

In aggregate, digital tools like these make a large difference. In the past, ignorance protected inefficient or lower-quality sellers from being unmasked by unsuspecting consumers, while geography limited competition from other sellers. With the introduction of structured comparison sites like FindTheBest.com and Kayak, airline travel, banking, insurance, car sales, motion pictures, and many other industries are being transformed by consumers' ability to search for and compare competing sellers. No longer can a seller of substandard services expect to feed on a continuing stream of naïve or ill-informed consumers. No longer can the seller expect to be insulated from competitors in other locations who can deliver a better service for less. Research by Michael Luca of Harvard Business School has found that the increased transparency has helped smaller independent restaurants compete with bigger chains because customers can more quickly find quality food via rating services like

* Yes, our long-time friend, the same Robert J. Gordon we discussed in chapter 6. See http://faculty-web.at.northwestern.edu/economics/gordon/p376_ipm_final_060313.pdf.

Yelp, reducing their reliance on brand names' expensive marketing campaigns.[17]

The intangible benefits delivered by the growing sharing economy—better matches, timeliness, customer service, and increased convenience—are exactly the types of benefits identified by the 1996 Boskin Commission as being poorly measured in our official price and GDP statistics.[18] This is another way in which our true growth is greater than the standard data suggest.

Intangible Assets

Just as free goods rather than physical products are an increasingly important share of consumption, intangibles also make up a growing share of the economy's capital assets. Production in the second machine age depends less on physical equipment and structures and more on the four categories of intangible assets: intellectual property, organizational capital, user-generated content, and human capital.

Intellectual property includes patents and copyrights. The rate of patenting by American inventors has been increasing rapidly since the 1980s,[19] and other types of intellectual assets have also grown.[20] In addition, a lot of research and development (R&D) is never formalized as intellectual property but is still very valuable.

The second—and even larger—category of intangibles is organizational capital like new business processes, techniques of production, organizational forms, and business models. Effective uses of the new technologies of the second machine age almost invariably require changes in the organization of work. For instance, when companies spend millions of dollars on computer hardware and software for a new enterprise resource planning system, they typically also include process changes that are three to five times as costly as the original investments in hardware and software. Yet, while the hardware and software spending generally shows up as additions to the nation's capital stock, the new business processes, which often outlast the hardware, are generally not counted as capital. Our research suggests that a correct accounting for computer-related intangible assets would add over $2 trillion to the official estimates of the capital assets in the United States economy.[21]

User-generated content is a smaller but rapidly growing third category of intangible assets. Users of Facebook, YouTube, Twitter, Instagram, Pinterest, and other types of online content not only consume this free content and gain the consumer surplus discussed above but also produce most of the content. There are 43,200 hours of new YouTube videos created each day,[22] as well as 250 million new photos uploaded each day on Facebook.[23] Users also contribute valuable but unmeasured content in the form of reviews on sites like Amazon, TripAdvisor, and Yelp. In addition, user-generated content includes the simple binary information used to sort reviews and present the best content first (e.g., when Amazon asks "Was this review helpful to you?"). Hardware and software companies now compete to improve the productivity of user-generated content activities. For example, smartphones and apps for smartphones now include easy or automatic tools for posting photos on Facebook. This content has value to other users and can be thought of as yet another type of intangible capital asset that is being added to our collective wealth.

The fourth and biggest category is the value of human capital. The many years that we all spend in schools learning skills like reading, writing, and arithmetic—as well as the additional learning that happens on the job and on our own—makes us more productive and, in some cases, is intrinsically rewarding. It is also a contribution to the nation's capital stock. According to Dale Jorgenson and Barbara Fraumeni, the value of human capital in the United States is five to ten times larger than the value of all the physical capital in the United States.[24] Human capital has not always been this important to the economy. The great economist Adam Smith understood that one of the great drawbacks of the first machine age was the way it forced workers to do repetitive tasks. In 1776, he noted, "The man whose whole life is spent in performing a few simple operations, of which the effects are perhaps always the same, or very nearly the same, has no occasion to exert his understanding."[25] As we'll discuss further later in the book, investments in human capital will be increasingly important as routine tasks become automated and the need for human creativity increases.

Important as these intangible assets are, the official GDP ignores them. User-generated content, for example, involves unmeasured labor creating an unmeasured asset that is consumed in unmeasured ways to create unmeasured consumer surplus. In recent years, however, there have been some efforts to create experimental 'satellite accounts.' They track some of these categories of intangible assets in the U.S. economy. For instance, the new satellite accounts created by the Bureau of Economic Analysis estimate that investment in R&D capital accounted for about 2.9 percent of GDP and has increased economic growth by about 0.2 percent per year between 1995 and 2004.[26]

It's hard to say exactly how large the bias is from miscounting all the types of intangible assets, but we are reasonably confident the official data underestimate their contribution.*

* Unlike unmeasured intangible consumption goods, the bad measures of intangible capital goods don't automatically bias official productivity statistics. On one hand, like all intangibles, intangible capital goods make the output numbers bigger. But at the same time, they are also used for production and thus make the input numbers bigger. In a steady state where both the input and output numbers are growing at the same rate, these two effects cancel out, so there is no bias in the productivity numbers, defined as output/input. Steady growth has been roughly true for some types of intangibles, such as the human capital assets that are created by education. But other categories—like computer-related organizational capital or the user-generated capital on digital content sites—appears to have been growing rapidly. For these categories of intangible assets, the official productivity numbers understate the true growth of the economy.

New Metrics Are Needed for the Second Machine Age

It's a fundamental principle of management: what gets measured gets done. Modern GDP accounting was certainly a huge step forward for economic progress. As Paul Samuelson and Bill Nordhaus put it, "While the GDP and the rest of the national income accounts may seem to be arcane concepts, they are truly among the great inventions of the twentieth century."[27]

But the rise in digital business innovation means we need innovation in our economic metrics. If we are looking at the wrong gauges, we will make the wrong decisions and get the wrong outputs. If we measure only tangibles, then we won't catch the intangibles that will make us better off. If we don't measure pollution *and* innovation, then we will get too much pollution and not enough innovation. Not everything that counts can be counted, and not everything that can be counted, counts.

As Nobel Prize winner Joe Stiglitz put it:

> The fact that GDP may be a poor measure of well-being, or even of market activity, has, of course, long been recognized. But changes in society and the economy may have heightened the problems, at the same time that advances in economics and statistical techniques may have provided opportunities to improve our metrics.[28]

The new metrics will differ both in conception and execution. We can build on some of the existing surveys and techniques researchers have been using. For instance, the human development index uses health and education statistics to fill in some of the gaps in official GDP statistics[29]; the multidimensional poverty index uses ten different indicators—such as nutrition, sanitation, and access to water—to assess well-being in developing countries.[30] Childhood death rates and other health indicators are recorded in periodic household surveys like the Demographic and Health Surveys.[31]

There are several promising projects in this area. Joe Stiglitz, Amartya Sen, and Jean-Paul Fitoussi have created a detailed guide for how we can do a comprehensive overhaul of our economic statistics.[32] Another promising project is the Social Progress Index that Michael Porter, Scott Stern, Roberto Loria, and their colleagues are developing.[33] In Bhutan, they've begun measuring "Gross National Happiness." There is also a long-running poll behind the Gallup-Healthways Well-Being Index.[34]

These are all important improvements, and we heartily support them. But the biggest opportunity is in using the tools of the second machine age itself: the extraordinary volume, variety, and timeliness of data available digitally. The Internet, mobile phones, embedded sensors in equipment, and a plethora of other sources are delivering data continuously. For instance, Roberto Rigobon and Alberto Cavallo measure online prices from around the world on a daily basis to create an inflation index that is far timelier and, in many cases, more reliable, than official data gathered via monthly surveys with much smaller samples.[35] Other economists are using satellite mapping of nighttime artificial light sources to estimate economic growth in different parts of the world, and assessing the frequency of Google searches to understand changes in unemployment and housing.[36] Harnessing this information will produce a quantum leap in our understanding of the economy, just as it has already changed marketing, manufacturing, finance, retailing, and virtually every other aspect of business decision-making.

As more data become available and as the economy continues to change, the ability to ask the right questions will become even more vital. No matter how bright the light is, you won't find your keys by searching under a lamppost if that's not where you lost them. We must think hard about what it is we really value, what we want more of, and what we want less of. GDP and productivity growth are important, but they are means to an end, not ends in and of themselves. Do we want to increase consumer surplus? Then lower prices or more leisure might be signs of progress, even if they result in a lower GDP. And, of course, many of our goals are nonmonetary. We shouldn't ignore the economic metrics, but neither should we let them crowd out our other values simply because they are more measurable.

In the meantime, we need to bear in mind that the GDP and productivity statistics overlook much of what we value, even when using a narrow economic lens. What's more, the gap between what we measure and what we value grows every time we gain access to a new good or service that never existed before, or when existing goods become free as they so often do when they are digitized.

CHAPTER 9

THE SPREAD

"An imbalance between rich and poor is the oldest and most fatal ailment of all republics."

—Plutarch

Of the 3.5 trillion photos that have been snapped since the first image of a busy Parisian street in 1838, fully 10 percent were taken in the last year.[1] Until recently, most photos were analog, created using silver halide and other chemicals. But analog photography peaked in 2000.[2] Today, over 2.5 billion people have digital cameras and the vast majority of photos are digital.[3] The effects are astonishing: it has been estimated that more photos are now taken every two minutes than in all of the nineteenth century.[4] We now record the people and events of our lives with unprecedented detail and frequency, and share them more widely and easily than ever before.

While digitization has obviously increased the quantity and convenience of photography, it has also profoundly changed the economics of photography production and distribution. A team of just fifteen people at Instagram created a simple app that over 130 million customers use to share some sixteen billion photos (and counting).[5] Within fifteen months of its founding, the company was sold for over $1 billion to Facebook. In turn, Facebook itself reached one billion users in 2012. It had about 4,600 employees[6] including barely 1,000 engineers.[7]

Contrast these figures with pre-digital behemoth Kodak, which also helped customers share billions of photos. Kodak employed 145,300 people at one point, one-third of them in Rochester, New York, while indirectly employing thousands more via the extensive supply chain and retail distribution channels required by companies in the first machine age. Kodak made its founder, George Eastman, a rich man, but it also provided middle-class jobs for generations of people and created a substantial share of the wealth created in the city of Rochester after company's founding in 1880. But 132 years later, a few months before Instagram was sold to Facebook, Kodak filed for bankruptcy.[8] Photography has never been more popular. Today, seventy billion photos are uploaded to Facebook each year, and many times more are shared via other digital services like Flickr at nearly zero cost. These photos are all digital, so hundreds of thousands of people who used to work making photography chemicals and paper are no longer needed. In a digital age, they need to find some other way to support themselves.

The evolution of photography illustrates *the bounty* of the second machine age, the first great economic consequence of the exponential, digital, combinatorial progress taking place at present. The second one, *spread*, means there are large and growing differences among people in income, wealth, and other important circumstances of life. We've created a cornucopia of images, sharing nearly four hundred billion "Kodak moments" each year with a few clicks of a mouse or taps on a screen. But companies like Instagram and Facebook employ a tiny fraction of the people that were needed at Kodak. Nonetheless, Facebook has a market value several times greater than Kodak ever did and has created at least seven billionaires so far, each of whom has a net worth ten times greater than George Eastman did. The shift from analog to digital has delivered a bounty of digital photos and other goods, but it has also contributed to an income distribution that is far more spread out than before.

Photography is not an isolated example of this shift. Similar stories have been and will be told in music and media; in finance and publishing; in retailing, distribution, services, and manufacturing. In almost every industry, technological progress will bring unprecedented bounty. More wealth will be created with less work. But at least in our current economic system, this progress will also have enormous effects on the distribution income and wealth. If the work a person produces in one hour can instead be produced by a machine for one dollar, then a profit-maximizing employer won't offer a wage for that job of more than one dollar. In a free-market system, either that worker must accept a wage of one dollar an hour or find some new way to make a living. Conversely, if a person finds a new way to leverage insights, talents, or skills across one million new customers using digital technologies, then he or she might earn one million times as much as would be possible otherwise. Both theory and data suggest that this combination of bounty and spread is not a coincidence. Advances in technology, especially digital technologies, are driving an unprecedented reallocation of wealth and income. Digital technologies can replicate valuable ideas, insights, and innovations at very low cost. This creates bounty for society and wealth for innovators, but diminishes the demand for previously important types of labor, which can leave many people with reduced incomes.

The combination of bounty and spread challenges two common though contradictory worldviews. One common view is that advances in technology always boost incomes. The other is that automation hurts workers' wages as people are replaced by machines. Both of these have a kernel of truth, but the reality is more subtle. Rapid advances in our digital tools are creating unprecedented wealth, but there is no economic law that says all workers, or even a majority of workers, will benefit from these advances.

For almost two hundred years, wages did increase alongside productivity. This created a sense of inevitability that technology helped (almost) everyone. But more recently, median wages have stopped tracking productivity, underscoring the fact that such a decoupling is not only a theoretical possibility but also an empirical fact in our current economy.

How's the Median Worker Doing?

Let's review some basic facts.

A good place to start is median income—the income of the person at the fiftieth percentile of the total distribution. The year 1999 was the peak year for the real (inflation-adjusted) income of the median American household. It reached $54,932 that year, but then started falling. By 2011, it had fallen nearly 10 percent to $50,054, even as overall GDP hit a record high. In particular, wages of unskilled workers in the United States and other advanced countries have trended downward.

Meanwhile, for the first time since before the Great Depression, over half the total income in the United States went to the top 10 percent of Americans in 2012. The top 1 percent earned over 22 percent of income, more than doubling their share since the early 1980s. The share of income going to the top hundredth of one percent of Americans, a few thousand people with annual incomes over $11 million, is now at 5.5 percent, after increasing more between 2011 and 2012 than any year since 1927–28.[9]

Several other metrics have also been increasingly unequal. For instance, while overall life expectancy continues to rise, life expectancies for some groups have started to fall. According to a study by S. Jay Olshansky and his colleagues published in *Health Affairs*, the average American white woman without a high school diploma had a life expectancy of 73.5 years in 2008, compared to 78.5 years in 1990. Life expectancy for white men without a high school education fell by three years during this period.[10]

It's no wonder that protests broke out across America even as it was beginning to recover from the Great Recession. The Tea Party movement on the right and the Occupy movement on the left each channeled the anger of the millions of Americans who felt the economy was not working for them. One group emphasized government mismanagement and the other abuses in the financial services sector.

How Technology Is Changing Economics

While undoubtedly both of these problems are important, the more fundamental challenge is deep and structural, and is the result of the diffusion to the second machine age technologies that increasingly drive the economy.

Recently we overheard a businessman speaking loudly (and cheerfully) into his mobile phone: "No way. I don't use an H&R Block tax preparer anymore. I've switched to TurboTax software. It's only forty-nine dollars, and it's much quicker and more accurate. I love it!" The businessman was better off. He had a better service at a lower price. Multiplied by millions of customers, TurboTax has created a great deal of value for its users, not all of which even shows up in the GDP statistics. The creators of TurboTax are also better off—one is a billionaire. But tens of thousands of tax preparers now find their jobs and incomes threatened.

The businessman's experience holds a mirror to the broader changes in the economy. Consumers are better off and enormous wealth is created, but a relatively small group of people often earns most of the income from the new products or services. Like the chemists who used silver halide to create camera film in the 1990s, human tax preparers have a hard time competing with machines. They can be made worse off by advances in technology, not just relative to the winners, but also relative to their income when they were working with the older technologies.

The crucial reality from the standpoint of economics is that it takes only a relatively small number of designers and engineers to create and update a program like TurboTax. As we saw in chapter 4, once the algorithms are digitized they can be replicated and delivered to millions of users at almost zero cost. As software moves to the core of every industry, this type of production process and this type of company increasingly populates the economy.

A Smaller Slice of a Bigger Pie

What happens when you scale up these types of examples to a whole economy? Is there something bigger going on? The data say yes.

Between 1983 and 2009, Americans became vastly wealthier overall as the total value of their assets increased. However, as noted by economists Ed Wolff and Sylvia Allegretto, the bottom 80 percent of the income distribution actually saw a net *decrease* in their wealth.[11] Taken as a group, the top 20 percent got not 100 percent of the increase, but more than 100 percent. Their gains included not only the trillions of dollars of wealth newly created in the economy but also some additional wealth that was shifted in their direction from the bottom 80 percent. The distribution was also highly skewed even among relatively wealthy people. The top 5 percent got 80 percent of the nation's wealth increase; the top 1 percent got over half of that, and so on for ever-finer subdivisions of the wealth distribution. In an oft-cited example, by 2010 the six heirs of Sam Walton's fortune, earned when he created Walmart, had more net wealth than the bottom 40 percent of the income distribution in America.[12] In part, this reflects the fact that thirteen million families had a negative net worth.

Along with wealth, the income distribution has also shifted. The top 1 percent increased their earnings by 278 percent between 1979 and 2007, compared to an increase of just 35 percent for those in the middle of the income distribution. The top 1 percent earned over 65 percent of income in the United States between 2002 and 2007. According to Forbes, the collective net worth of the wealthiest four hundred Americans reached a record two trillion dollars in 2013, more than doubling since 2003.[13]

IN SHORT, median income has increased very little since 1979, and it has actually fallen since 1999. But that's not because growth of overall income or productivity in America has stagnated; as we saw in chapter 7, GDP and productivity have been on impressive trajectories. Instead, the trend reflects a significant reallocation of who is capturing the benefits of this growth, and who isn't.

This is perhaps easiest to see if one compares *average* income with *median* income. Normally, changes in the average income (total income divided by the total number of people) are not very different from changes in median income (income of the person exactly in the middle of the income distribution—half earn more and half earn less). However, in recent years, the trends have diverged significantly, as shown in figure 9.1.

How is this possible? Consider a simple example. Ten bank tellers are drinking beers at a bar. Each of them makes $30,000 a year, so both the mean and median income of this group is $30,000. In walks the CEO and orders a beer. Now the average income of the group has skyrocketed, but the median hasn't changed at all. In general, the more skewed the incomes, the more the mean tends to diverge from the median. This is what has happened not only in our hypothetical bar but also in America as a whole.

Overall, between 1973 and 2011, the median hourly wage barely changed, growing by just 0.1 percent per year. In contrast, as discussed in chapter 7, productivity grew at an average of 1.56 percent per year during this period, accelerating a bit to 1.88 percent per year from 2000 to 2011. Most of the growth in productivity directly translated into comparable growth in average income. The reason why median income growth was so much lower was primarily because of increases in inequality.[14]

FIGURE 9.1 Real GDP vs. Median Income per Capita

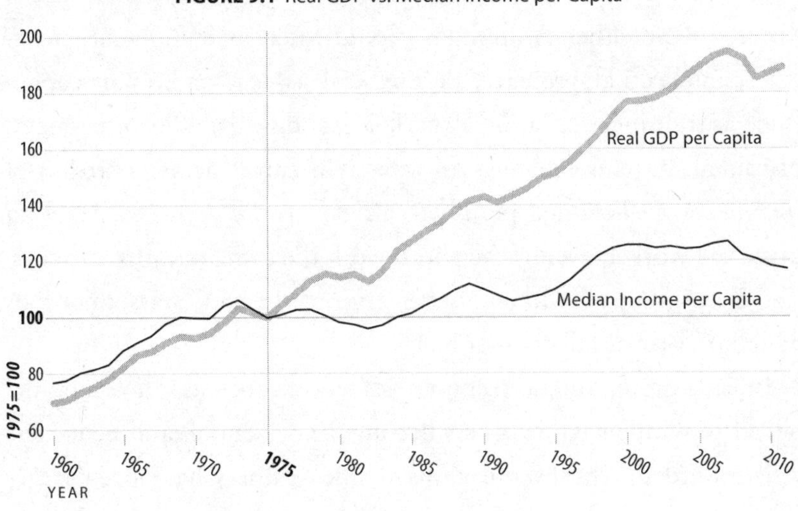

The Three Pairs of Winners and Losers

In the past couple of decades, we've seen changes in tax policy, greater overseas competition, ongoing government waste, and Wall Street shenanigans. But when we look at the data and research, we conclude that none of these are the primary driver of growing inequality. Instead, the main driver is exponential, digital, and combinatorial change in the technology that undergirds our economic system. This conclusion is bolstered by the fact that similar trends are apparent in most advanced countries. For instance, in Sweden, Finland, and Germany, income inequality has actually grown more quickly over the past twenty to thirty years than in the United States.[15] Because these countries started with less inequality in their income distributions, they continued to be less unequal than the United States, but the underlying trend is similar worldwide across sometimes markedly different institutions, government policies, and cultures.

As we discussed in our earlier book *Race Against the Machine*, these structural economic changes have created three overlapping pairs of winners and losers. As a result, not everyone's share of the economic pie is growing. The first two sets of winners are those who have accumulated significant quantities of the right capital assets. These can be either nonhuman capital (such as equipment, structures, intellectual property, or financial assets), or human capital (such as training, education, experience, and skills). Like other forms of capital, human capital is an asset that can generate a stream of income. A well-trained plumber can earn more each year than an unskilled worker, even if they both work the same number of hours. The third group of winners is made up of the superstars among us who have special talents—or luck.

In each group, digital technologies tend to increase the economic payoff to winners while others become less essential, and hence less well rewarded. The overall gains to the winners have been larger than total losses for everyone else. That simply reflects the fact we discussed earlier: productivity and total income have grown in the overall economy. This good news offers little consolation to those who are falling behind. In some cases the gains, however large, have been concentrated among a relatively small group of winners, leaving the majority of people worse off than before.

Skill-Biased Technical Change

The most basic model economists use to explain technology's impact treats it as a simple multiplier on everything else, increasing overall productivity evenly for everyone.[16] This model can be described in mathematical equations. It is used in most introductory economics classes and provides the foundation for the common—and until recently, very sensible—intuition that a rising tide of technical progress will lift all boats, that it will make all workers more productive and hence more valuable. With technology as a multiplier, an economy is able to produce more output each year with the same inputs, including labor. And in the basic model all labor is affected equally by technology, meaning every hour worked produces more value than it used to.

A slightly more complex model allows for the possibility that technology may not affect all inputs equally, but rather may be 'biased' toward some and against others. In particular, in recent years, technologies like payroll processing software, factory automation, computer-controlled machines, automated inventory control, and word processing have been deployed for routine work, *substituting* for workers in clerical tasks, on the factory floor, and doing rote information processing.

By contrast, technologies like big data and analytics, high-speed communications, and rapid prototyping have *augmented* the contributions made by more abstract and data-driven reasoning, and in turn have increased the value of people with the right engineering, creative, or design skills. The net effect has been to decrease demand for less skilled labor while increasing the demand for skilled labor. Economists including David Autor, Lawrence Katz and Alan

FIGURE 9.2 Wages for Full-Time, Full-Year Male U.S. Workers, 1963–2008

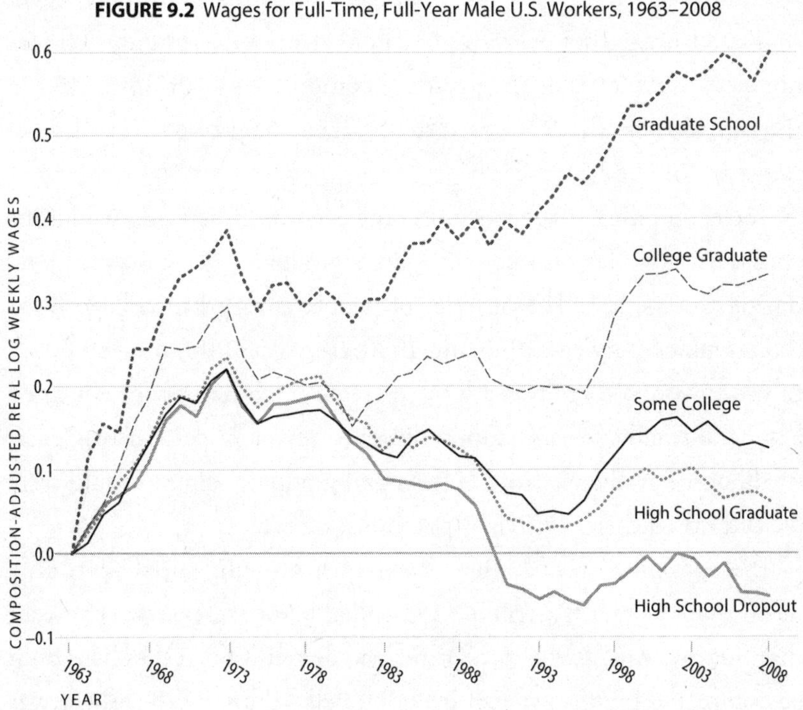

Krueger, Frank Levy and Richard Murnane, Daron Acemoglu, and many others have documented this trend in dozens of careful studies.[17] They call it *skill-biased technical change.* By definition, skill-biased technical change favors people with more human capital.

The effects of skill-biased technical change can be vividly seen in figure 9.2, which is based on data from a paper by MIT economists Daron Acemoglu and David Autor.[18] The lines tell a story about the diverging paths of millions of workers over recent generations. Before 1973, American workers all enjoyed brisk wage growth. The rising tide of productivity increased everyone's incomes, regardless of their educational levels. Then came the massive oil shock and recession of the 1970s, which reversed the gains for all groups. However, after that, we began to see a growing spread of incomes. By the early 1980s, those with college degrees started to see their wages growing again. Workers with graduate degrees did particularly well. Meanwhile, workers without college degrees were confronted with a much less attractive labor market. Their wages stagnated or, if they were high school dropouts, actually fell. It's not a coincidence that the personal computer revolution started in the early 1980s; the PC was actually *Time* magazine's "machine of the year" in 1982.

The economics of the story become even more striking when one considers that the number of college graduates grew very rapidly during this period. The number of people enrolled in college more than doubled between 1960 and 1980, from 758,000 to 1,589,000.[19] In other words, there was a large increase in the supply of educated labor. Normally, greater supply leads to lower prices. In this case, the flood of graduates from college and graduate school should have pushed down their relative wages, but it didn't.

The combination of higher pay despite growing supply can only mean that the relative *demand* for skilled labor increased even faster than supply. And at the same time, the demand for tasks that could be completed by high school dropouts fell so rapidly that there was a glut of this type of worker, even though their ranks were thinning. The lack of demand for unskilled workers meant ever-lower wages for those who continued to compete for low-skill jobs. And because most of the people with the least education already had the lowest wages, this change increased overall income inequality.

Organizational Coinvention

While a one-for-one substitution of machines for people sometimes occurs, a broader reorganization in business culture may have been an even more important path for skill-biased change. Work that Erik did with Stanford's Tim Bresnahan, Wharton's Lorin Hitt, and MIT's Shinkyu Yang found that companies used digital technologies to reorganize decision-making authority, incentives systems, information flows, hiring systems, and other aspects of their management and organizational processes.[20] This coinvention of organization and technology not only significantly increased productivity but tended to require more educated workers and reduce demand for less-skilled workers. This reorganization of production affected those who worked directly with computers as well as workers who, at first glance, seemed to be far from the technology. For instance, a designer with a knack for style might find herself in greater demand at a company with flexible equipment in distant factories that can quickly adapt to the latest fashions, while an airport ticket agent might find himself replaced by an Internet website he never knew existed, let alone worked with.

Among the industries in the study, each dollar of computer capital was often the catalyst for more than ten dollars of complementary investments in "organizational capital," or investments in training, hiring, and business process redesign.[21] The reorganization often eliminates a lot of routine work, such as repetitive order entry, leaving behind a residual set of tasks that require relatively more judgment, skills, and training.

Companies with the biggest IT investments typically made the biggest organizational changes, usually with a lag of five to seven years before seeing the full performance benefits.[22] These companies had the biggest increase in the demand for skilled work relative to unskilled work.[23] The lags reflected the time that it takes for managers and workers to figure out new ways to use the technology. As we saw in our earlier discussion of electrification and factory design, businesses rarely get significant performance gains from simply "paving the cowpaths" as opposed to rethinking how the business can be redesigned to take advantage of new technologies.[24] Creativity and organizational redesign are crucial to investments in digital technologies.*

This means that the best way to use new technologies is usually not to make a literal substitution of a machine for each human worker, but to restructure the process. Nonetheless, some workers (usually the less skilled ones) are still eliminated from the production process and others are augmented (usually those with more education and training), with predictable effects on the wage structure. Compared to simply automating existing tasks, this kind of organizational coinvention requires more creativity on the part of entrepreneurs, managers, and workers, and for that reason it tends to take time to implement the changes after the initial invention and introduction of new technologies. But once the changes are in place, they generate the lion's share of productivity improvements.

The Skill Set Affected by Computerization Is Evolving

If we look more closely at the jobs eliminated as companies reorganized, *skill-biased technical change* can be a bit of a misleading moniker.

* This echoes the productivity effects of electricity discussed earlier. As with digital technologies, the biggest gains did not occur until factories were redesigned, and even workers who didn't work directly with the new machines were significantly affected.

In particular, it would be a mistake to assume that all 'college-level tasks' are hard to automate while 'kindergarten tasks' are easy. In recent years, 'low-skill tasks' haven't always been the ones being automated; more often it has been 'tasks that machines can do better than humans.' Of course, that's a bit of a tautology, but a useful tautology nonetheless. Repetitive work on an assembly line is easier to automate than the work of a janitor. Routine clerical work like processing payments is easier to automate than handling customers' questions. At present, machines are not very good at walking up stairs, picking up a paperclip from the floor, or reading the emotional cues of a frustrated customer.

To capture these distinctions, work by our MIT colleagues Daron Acemoglu and David Autor suggests that work can be divided into a two-by-two matrix: cognitive versus manual and routine versus nonroutine.[25] They found that the demand for work has been falling most dramatically for routine tasks, regardless of whether they are cognitive or manual. This leads to job polarization: a collapse in demand for middle-income jobs, while nonroutine cognitive jobs (such as financial analysis) and nonroutine manual jobs (like hairdressing) have held up relatively well.

Building on Acemoglu and Autor's work, economists Nir Jaimovich of Duke University and Henry Siu of the University of British Columbia found a link between job polarization and the jobless recoveries that have defined the last three recessions. For most of the nineteenth and twentieth centuries, employment usually rebounded strongly after each recession, but since the 1990s employment didn't recover briskly after recessions. Again, it's not a coincidence that as the computerization of the economy advanced, post-recession hiring patterns changed. When Jaimovich and Siu compared the 1980s, 1990s, and 2000s, they found that the demand for routine cognitive tasks such as cashiers, mail clerks, and bank tellers and routine manual tasks such as machine operators, cement masons, and dressmakers was not only falling, but falling at an accelerating rate. These jobs fell by 5.6 percent between 1981 and 1991, 6.6 percent between 1991 and 2001, and 11 percent between 2001 and 2011.[26] In contrast, both nonroutine cognitive work and nonroutine manual work grew in all three decades.

CONVERSATIONS WITH senior executives help explain this pattern in the data. A few years ago, we had a very candid discussion with one CEO, and he explained that he knew for over a decade that advances in information technology had rendered many routine information-processing jobs superfluous. At the same time, when profits and revenues are on the rise, it can be hard to eliminate jobs. When the recession came, business as usual obviously was not sustainable, which made it easier to implement a round of painful streamlining and layoffs. As the recession ended and profits and demand returned, the jobs doing routine work were not restored. Like so many other companies in recent years, his organization found it could use technology to scale up without these workers.

As we saw in chapter 2, this reflects Moravac's paradox, the insight that the sensory and motor skills we use in our everyday lives require enormous computation and sophistication.[27] Over millions of years, evolution has endowed us with billions of neurons devoted to the subtleties of recognizing a friend's face, distinguishing different types of sounds, and using fine motor control. In contrast, the abstract reasoning that we associate with 'higher thought' like arithmetic or logic is a relatively recent skill, developed over only a few thousand years. It often requires simpler software and less computer power to mimic or even exceed human capabilities on these types of tasks.

Of course, as we've seen throughout this book, the set of tasks machines can do is not fixed. It is constantly evolving, just as our use of the word "computer" itself has evolved from referring to a job that humans do to referring to a piece of equipment.

In the early 1950s, machines were taught how to play checkers and could soon beat respectable amateurs.[28] In January 1956, Herbert Simon returned to teaching his class and told his students, "Over Christmas, Al Newell and I invented a thinking machine." Three years later, they created a computer program modestly called the "General Problem Solver," which was designed to solve, in principle, any logic problem that could be described by a set of formal rules. It worked well on simple problems like Tic-Tac-Toe or the slightly harder Tower of Hanoi puzzle, although it didn't scale up to most real-world problems because of the combinatorial explosion of possible options to consider.

Cheered by their early successes and those of other artificial intelligence pioneers like Marvin Minsky, John McCarthy and Claude Shannon, and Simon and Newell were quite optimistic about how rapidly machines would master human skills, predicting in 1958 that a digital computer would be the world chess champion by 1968.[29] In 1965, Simon went so far as to predict, "machines will be capable, within twenty years, of doing any work a man can do."[30]

Simon won the Nobel Prize in Economics in 1978, but he was wrong about chess, not to mention all the other tasks that humans can do. His mistake may have been more about the timing than the ultimate outcome. After Simon made his prediction, computer chess programs improved by about forty points per year on the official Elo chess rating system. On May 11, 1997, forty years after Simon's prediction, an IBM computer called Deep Blue beat the world chess champion, Gary Kasparov, after a six-game match. Today, no human can beat even a mid-tier computer chess program. In fact, software and hardware have progressed so rapidly that by 2009, chess programs running on ordinary personal computers, and even mobile phones, have achieved grandmaster levels with Elo ratings of 2,898 and have won tournaments against the top human players.[31]

Labor and Capital

Technology is not only creating winners and losers among those with differing amounts of human capital, it is also changing the way national income is divided between the owners of physical capital and labor (people like factory owners and factory workers)—the two classical inputs to production.

When Terry Gou, the founder of Foxconn, purchased thirty thousand robots to work in the company's factories in China, he was substituting capital for labor.[32] Similarly, when an automated voice-response system usurps some of the functions of human call center operators, the production process has more capital and less labor. Entrepreneurs and managers are constantly making these types of decisions, weighing the relative costs of each type of input, as well as the effects on the quality, reliability, and variety of output that can be produced.

Rod Brooks estimates that the Baxter robot we met in chapter 2 works for the equivalent of about four dollars per hour, including all costs.[33] As we discussed at the start of this chapter, to the extent that a factory owner previously employed a human to do the same task that Baxter could do, the economic incentive would be to substitute capital (Baxter) for labor as long as the human was paid more than four dollars per hour. If output stays the same, and assuming no new hires are made in engineering, management, or sales at the company, it would increase the ratio of capital to labor input.*

Compensation of the remaining workers could go up or down in the wake of Baxter's arrival. If their work is a close substitute for the robots', then there will be downward pressure on human wages.

* The effect on the economy overall would depend on how other companies reacted. Output would likely increase at companies that design and build robots and, depending on how capital-intensive they are, the net ratio of capital to labor in the overall economy could increase, decrease, or stay the same. We'll discuss these effects in more detail in chapter 12.

That will grow even worse if Moore's Law and other advances allow future versions of Baxter to work for two dollars per hour, and then one dollar per hour, and so on, while handling an increasing variety and complexity of tasks. However, economic theory also holds open the possibility that the remaining workers would see an increase in pay. In particular, if their work complements the technology, then demand for their services will increase. In addition, as technical advances increase labor productivity, employers can afford to pay more for each worker. In some cases, this is reflected directly in higher wages and benefits. In other cases, the prices of products and services fall, so the real wage of workers increases as they are able to buy more with each dollar. As productivity improves, total amount of output per person would increase but the amount earned by human workers could either fall or rise, with the remainder going to capital owners.

Of course, almost every economy has been using technology to substitute capital for labor for decades, if not centuries. Automatic threshing machines replaced a full 30 percent of the agricultural labor force in the middle of the nineteenth century, and industrialization continued at a brisk pace throughout the twentieth century. Nineteenth-century economists like Karl Marx and David Ricardo predicted that the mechanization of the economy would worsen the fate of workers, ultimately driving them to a subsistence wage.[34]

What has actually happened to the relative share of capital and labor? Historically, despite changes in the technology of production, the share of overall GDP going to labor has been surprisingly stable, at least until recently. As a result, wages and living standards have grown dramatically, roughly in line with the dramatic increases in productivity. In part, this reflects the increases in human capital that have paralleled the more visible increases in equipment and buildings in the economy. Dale Jorgenson and his colleagues have estimated that the overall magnitude of the human capital in the U.S. economy, as measured by its economic value, is as much as ten times the value of the physical capital.[35] As a result, labor compensation has grown along with payments to owners of physical capital via profits, dividends, and capital gains.

Figure 9.3 shows that in the past decade, the relatively consistent division between the shares of income going to labor and physical capital seems to be coming to an end. As noted by Susan Fleck, John Glaser, and Shawn Sprague in the *Monthly Labor Review*: "Labor share averaged 64.3 percent from 1947 to 2000. In the United States, the share of GDP going to labor has declined over the past decade, falling to its lowest point in the third quarter of 2010, 57.8 percent."[36] What's more, this is a global phenomenon. Economists Loukas Karabarbounis and Brent Neiman of the University of Chicago find that "the global labor share has significantly declined since the early 1980s, with the decline occurring within the large majority of countries and industries."[37] They argue that this decline is likely due to the technologies of the information age.

The fall in labor's share is in part the consequence of two trends

FIGURE 9.3 Wage Share of GDP vs. Corporate Profit Share of GDP

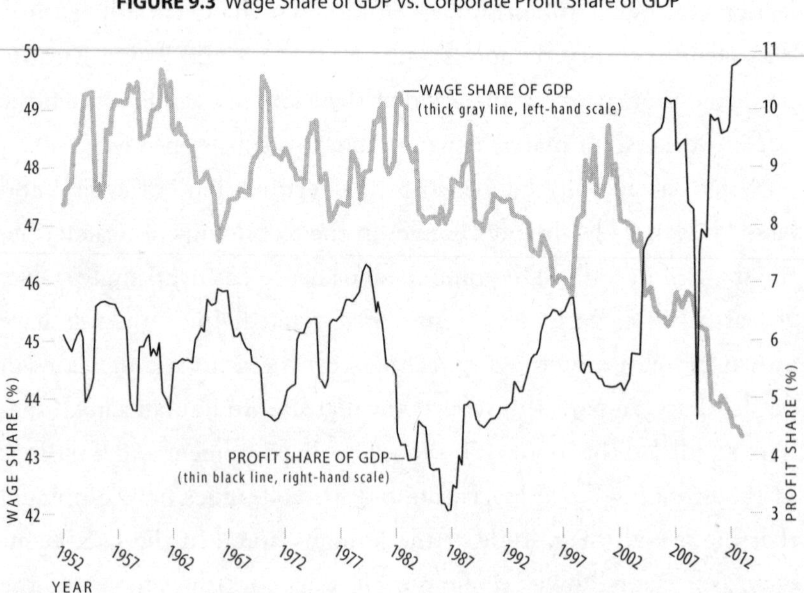

we have already noted: fewer people are working, and wages for those who are working are lower than before. As a result, while labor compensation and productivity in the past rose in tandem, in recent years a growing gap has opened.

If productivity is growing and labor as a whole isn't capturing the value, who is? Owners of physical capital, to a large extent. While the economy remained mired in a slump, profits reached historic highs last year, both in absolute terms ($1.6 trillion) and as a share of GDP (26.2 percent in 2010, up from the 1960–2007 average of 20.5 percent).[38] Meanwhile, real spending on capital equipment and software has soared by 26 percent while payrolls have remained essentially flat, as noted by Kathleen Madigan.[39]

What's more, the collapse in the share of GDP going to labor actually understates how the situation has deteriorated for the typical worker. The official measure of labor compensation includes soaring wages for a small number of superstars in media, finance, sports, and corporate positions. Furthermore, it is debatable that all of the compensation going to CEOs and other top executives is solely due to their 'labor' income. It may also reflect their bargaining power, as suggested by Harvard Law Professor Lucian Bebchuk and others.[40] In this sense, it might make sense to think of CEOs' income as due to their control of capital, not labor, at least in part.

While the share of national income to capital has been growing at the expense of labor, economic theory does not necessarily predict that this will continue, even if robots and other machines take over more and more work. The threat to capital's share comes not (just) from the bargaining power of various types of human labor, from CEOs or labor unions but, ironically, from other capital. In a free market, the biggest premiums go to the scarcest inputs needed for production. In a world where capital can be replicated at a relatively low cost (think of computer chips or even software), the marginal value of capital will tend to fall, even if more capital is used overall. The value of existing capital will actually be driven down as new capital is added cheaply at the margin. Thus, the rewards earned by capitalists may not automatically grow relative to labor. Instead the shares will depend on the exact details of the production, distribution, and governance systems.

Most of all, the payoff will depend on which inputs to production are scarcest. If digital technologies create cheap substitutes for labor, then it's not a good time to be a laborer. But if digital technologies also can increasingly substitute for capital, then capital owners shouldn't expect to earn high returns either. What will be the scarcest, and hence the most valuable, resource in the second machine age? This question brings us to our next set of winners and losers: superstars versus everyone else.

CHAPTER 10

THE BIGGEST WINNERS: STARS AND SUPERSTARS

"One machine can do the work of fifty ordinary men. No machine can do the work of one extraordinary man."

—Elbert Hubbard

WE'VE SEEN THAT SKILL-BIASED technical change has increased the relative demand for highly educated workers while reducing demand for less educated workers whose jobs frequently involve routine cognitive and manual tasks. In addition, capital-biased technological changes that encourage substitution of physical capital for labor have increased the profits earned by capital owners and reduced the share of income going to labor. In each case, historic amounts of wealth have been created. In each case, we also have seen increases in the earnings of the winners relative to the losers. But the biggest changes of all are driven by a third gap between winners and losers: the gap between the superstars in a field and everyone else.

Mind the Gap

Call it talent-biased technical change.* In many industries, the difference in payout between number one and second-best has widened into a canyon. As a controversial Nike ad noted, you don't win silver, you lose gold.[1] When 'winner-take-all' markets become more important, income inequality will rise because pay at the very top pulls away from pay in the middle.[2]

The growing gaps in wages between people with and without college education, and between capital owners and workers, have been

* If you're a cynic, you might call it luck-biased technical change.

dwarfed by even bigger changes at the very top. As noted earlier, between 2002 and 2007, the top 1 percent got two-thirds of all the profits from the growth in the U.S. economy. But who are the 1 percent? They aren't all on Wall Street. University of Chicago economist Steve Kaplan found that most of them are in other industries: in media and entertainment, sports, and law—or they are entrepreneurs and senior executives.

If the top 1 percent are stars of a sort, they can look up to superstars who have seen even bigger increases. While the top 1 percent earned about 19 percent of all income in the United States, the top 1 percent of the 1 percent (or the top 0.01 percent)—saw their share of national income double from 3 percent to 6 percent between 1995 and 2007. This is nearly six times as much as the 0.01 percent earned between World War II and the late 1970s. In other words, the top 0.01 percent now get a bigger share of the top 1 percent of income than the top 1 percent get of the whole economy. Because it is hard to maintain anonymity when reporting data for small numbers of people, it is hard to get reliable data at income levels higher than the top 0.01 percent. After all, while there are over 1.35 million households in the top 1 percent with an average income of $1.12 million, the 0.01 percent represents just 14,588 families each with incomes over $11,477,000.*[3] But the evidence suggests that the spread of incomes continues at high levels of income with a fractal-like quality, with each subset of superstars watching an even smaller group of super-duper-stars pulling away.†

* In 2011, families with incomes above $367,000 were in the top 1 percent in the United States, but of course, the average reflects people with much higher incomes than that. See http://elsa.berkeley.edu/~saez/saez-US-topincomes-2011.pdf

† This is a characteristic of Power Law distributions, which we'll discuss later in this chapter.

How Superstars Thrive in the Winner-Take-All Economy

In the previous chapter, we saw Intuit's TurboTax automate the job of tax preparation, allowing a machine to do the jobs of hundreds of thousands of human tax preparers. That's an example of technology automating routine information-processing jobs, and also an example of capital substituting for labor. But most importantly, it's an example of the superstar economy in action. Intuit's CEO made $4 million last year and Scott Cook, the founder, is a billionaire.[4] Likewise, the fifteen people who created Instagram didn't need a lot of unskilled human helpers and did leverage some valuable physical capital. But most of all, they benefitted from their talent, timing, and ties to the right people.

Top performers in other industries have also seen their fortunes rise. J. K. Rowling, author of the *Harry Potter* series, is the world's first billionaire author in an industry not known for minting the super wealthy. As George Mason University's Alex Tabarrok notes of Rowling's success:

> Homer, Shakespeare and Tolkien all earned much less. Why? Consider Homer, he told great stories but he could earn no more in a night than say 50 people might pay for an evening's entertainment. Shakespeare did a little better. The Globe theater could hold 3000 and unlike Homer, Shakespeare didn't have to be at the theater to earn. Shakespeare's words were leveraged.[5]

J. R. R. Tolkien's words were leveraged further. By selling books, Tolkien could sell to hundreds of thousands, even millions of buyers in a year—more than have ever seen a Shakespeare play in four hundred years. And books were cheaper to produce than actors, which meant that Tolkien could earn a greater share of the revenues than did Shakespeare.

Technology has supercharged the ability of authors like Rowling to leverage their talents via digitization and globalization. Rowling's stories can be captured in movies and video games as well as text, but each of those formats, including the original books, can be transmitted globally at trivial cost. She and other superstar storytellers now reach billions of customers through a variety of channels and formats.

More often than not, when improvements in digital technologies make it more and more attractive to digitize something, superstars in various markets see a boost in their incomes while second-bests have a harder time competing. The top performers in music, sports, and other areas have also seen their reach and incomes grow since the 1980s.[6]

At the same time, others working in the content and entertainment industries have not seen a big increase. Only 4 percent of software developers in the burgeoning app economy have made over a million dollars.[7] Three-quarters of them made less than thirty thousand dollars. While a handful of writers, actors, or baseball players can become millionaires, many others struggle to make ends meet. A gold-medal winner at the Olympics can earn millions of dollars in endorsements, while the silver medal winner—let alone the person who placed tenth or thirtieth—is quickly forgotten, even if the difference is measured in tenths of a second and could have resulted from a gust of wind or a lucky bounce of the ball.

Even top executives have started earning rock-star compensation. The ratio of CEO pay to average worker pay increased from seventy in 1990 to three hundred in 2005. Much of this growth is linked to the greater use of information technology, according to research that Erik completed with his student Heekyung Kim.[8] One rationale for this increase in executive pay is that technology increases the reach, scale, or monitoring capacity of a decision-maker. If executives use digital technologies to observe activities in factories throughout the world, to give specific instructions for changing a process, and to make sure instructions are carried out with high fidelity, then the value of those decision-makers increases. Direct management via digital technologies makes a good manager more valuable than in earlier times when managers had diffuse control via long chains of subordinates, or when they could only affect a smaller scale of activities.

Direct digital oversight also makes hiring the best candidate rather than the second-best that much more important. Companies are ready to pay a premium for executives whom they perceive to be the best, reasoning that even a small difference in quality can have huge consequences for shareholders. The bigger the market value of a company, the more compelling the argument for trying to get the very best executive.[9] A single decision that increases value by a modest 1 percent is worth $100 million to a ten-billion-dollar company.

In a competitive market, even a small difference in the perceived talents of CEO candidates can lead to fairly large differences in their compensation. As economists Robert Frank and Philip Cook note in their book, *The Winner-Take-All Society*, "When a sergeant makes a mistake only the platoon suffers, but when a general makes a mistake the whole army suffers."[10]

When Relative Advantage Leads to Absolute Domination

The economics of superstars was first formally analyzed in 1981 by economist Sherwin Rosen.[11] In many markets, buyers with a choice among products or services will prefer the one with the best quality. When there are capacity constraints or significant transportation costs, then the best seller will only be able to satisfy a small fraction of the global market (for instance, in the 1800s, even the best singers and actors might perform for at most a few thousand people each year). Other inferior sellers will also have a market for their products. But what if a technology arises that lets each seller cheaply replicate his or her services and deliver them globally at little or no cost? Suddenly the top-quality provider can capture the whole market. The next-best provider might be almost as good, but it will not matter. Each time a market becomes more digital, these winner-take-all economics become a little more compelling.

Winner-take-all markets were just coming to the fore in the 1990s, when Frank and Cook wrote their remarkably prescient book. They compared these winner-take-all markets, where the compensation was mainly determined by *relative* performance, to traditional markets, where revenues more closely tracked *absolute* performance. To understand the distinction, suppose the best, hardest-working construction worker could lay one thousand bricks in a day while the tenth-best laid nine hundred bricks per day. In a well-functioning market, pay would reflect this difference proportionately, whether it could be attributed to more efficiency and skill, or simply to more hours of work. In a traditional market, someone who is 90 percent as skilled or works 90 percent as hard creates 90 percent as much value and thus can earn 90 percent as much money. That's absolute performance.

By contrast, a software programmer who writes a slightly better mapping application—one that loads a little faster, has slightly more complete data, or prettier icons—might completely dominate a market. There would likely be little, if any, demand for the tenth-best mapping application, even it got the job done almost as well. This is relative performance. People will not spend time or effort on the tenth-best product when they have access to the best. And this is not a case where quantity can make up for quality: ten mediocre mapping tools are no substitute for one good one. When consumers care mostly about relative performance, even a small difference in skill or effort or luck can lead to a thousand-fold or million-fold difference in earnings. There were a lot of traffic apps in the marketplace in 2013, but Google only judged one, Waze, worth buying for over one billion dollars.[12]

Why Winner-Take-All Is Winning

Why are winner-take-all markets more common now? Shifts in the technology for production and distribution, particularly these three changes:

a) the digitization of more and more information, goods, and services,

b) the vast improvements in telecommunications and, to a lesser extent, transportation, and

c) the increased importance of networks and standards.

Albert Einstein once said that black holes are where God divided by zero, and that created some strange physics. While the marginal costs of digital goods do not quite approach zero, they are close enough to create some pretty strange economics. As discussed in chapter 3, digital goods have much lower marginal costs of production than physical goods. Bits are cheaper than atoms, not to mention human labor.

Digitization creates winner-take-all markets because, as noted above, with digital goods capacity constraints become increasingly irrelevant. A single producer with a website can, in principle, fill the demand from millions or even billions of customers. Jenna Marbles's homemade YouTube video "How to trick people into thinking you're good looking," to take one wildly successful example, garnered 5.3 million views the week she posted it in July 2010.[13] She's now earned millions of dollars from over one billion viewings of her videos around the world. Every digital app developer, no matter how humble its offices or how small its staff, almost automatically becomes a micro-multinational, reaching global audiences with a speed that would have been inconceivable in the first machine age.

In contrast, the economics of personal services (nursing) or physical work (gardening) are very different, since each provider, no matter how skilled or hard-working, can only fulfill a tiny fraction of the overall market demand. When an activity transitions from the second category to the first the way tax preparation did, the economics shift toward winner-take-all outcomes. What's more, lowering prices, the traditional refuge for second-tier products, is of little benefit for anyone whose quality is not already at or near the world's best. Digital goods have enormous economies of scale, giving the market leader a huge cost advantage and room to beat the price of any competitor while still making a good profit.[14] Once their fixed costs are covered, each marginal unit produced costs very little to deliver.[15]

Improvements in Telecommunications: Reach Out and Touch More People

Secondly, winner-take-all markets have also been boosted by technological improvements in telecommunications and transportation that also expand the market individuals and companies can reach. When there are many small local markets, there can be a 'best' provider in each, and these local heroes frequently can all earn a good income. If these markets merge into a single global market, top performers have an opportunity to win more customers, while the next-best performers face harsher competition from all directions. A similar dynamic comes into play when technologies like Google or even Amazon's recommendation engine reduce search costs. Suddenly second-rate producers can no longer count on consumer ignorance or geographic barriers to protect their margins.

Digital technologies have aided the transition to winner-take-all markets, even for products we wouldn't think would have superstar status. In a traditional camera store, cameras typically are not ranked number one versus number ten. But online retailers make it easy to list products in rank order by customer ratings, or to filter results to include only products with every conceivable desirable feature. Products with lower rankings or only nine out of ten desirable features receive disproportionately lower sales from even small differences in quality, convenience, or pricing performance.[16]

Digital ranking and filtering create disproportional returns even in labor markets for workaday, non-superstar careers. Companies have digitized their hiring processes and use automated filters to winnow the flood of applicants. For example, companies can readily cull all the candidates that don't have a college degree as a simple expedient even if the job does not actually require a college education.[17] This can amplify a trickle of skill-biased technical change into a torrent of stardom for a lucky few. Similarly, job candidate resumes that miss the buzzword requirements might drop from consideration even if the 90-percent-qualified candidate might otherwise be a stellar employee.

Networks and Standards: The Value of Scale

Thirdly, the increased importance of networks (like the Internet or credit card networks) and interoperable products (like computer components) can also create winner-take-all markets. Just as low marginal costs create economies of scale on the production side, networks can create 'demand side economies of scale' that economists sometimes call *network effects*. We see them at work when users prefer products or services that other people are flocking to. If your friends keep in touch via Facebook, that makes Facebook more attractive to you, too. If you then join Facebook, the site becomes more valuable to your friends as well.

Sometimes network effects are indirect. You can make a phone call equally well to someone using an iPhone or an Android phone. But the total number of users on a given platform influences app developers: the bigger network of users will tend to attract more developers, or encourage app developers to invest more in a given platform. The more apps available for a given phone, the greater its appeal to users. Thus, your benefits from buying one or the other will be affected by the number of other users who buy the same product. When Apple's app ecosystem is strong, buyers will want to buy into that platform, attracting even more developers. But the opposite dynamic can unravel a dominant standard, as it almost did for the Apple Macintosh platform in the mid-1990s. Like low marginal costs, network effects can create both winner-take-all markets and high turbulence.[18]

The Social Acceptability of Superstars

In addition to the technical changes that have increased digitization, telecommunication, networks, and other factors that create superstar products and companies, there are more aspects at work in boosting superstar compensation for individuals. In some cases, cultural barriers to very large pay packages have fallen. CEOs, financial executives, actors, and professional athletes may be more willing to demand seven- or even eight-figure compensation deals. As more people get those deals, a positive feedback loop emerges: it becomes easier for others to make similar requests.

In fact, the concentration of wealth itself can create what Frank and Cook call "deep pocket" winner-take-all markets. As the great economist Alfred Marshall noted, "a rich client whose reputation or fortune or both are at stake will scarcely count any price too high to secure the services of the best man he can get."[19] If mass-market media enables an athlete like O. J. Simpson to earn millions, then he can afford to pay a lawyer like Alan Dershowitz millions to defend him in court, even if Dershowitz's services are not replicated to millions of people like Simpson's are. In a sense, Dershowitz is a superstar by proxy: he benefits from the ability of his superstar clients whose labor has been more directly leveraged by digitization and networks.*

* At least in his capacity as a courtroom lawyer. As an author or TV celeb-

Laws and institutions have also changed in ways that often boost the incomes of superstars. The top marginal tax rate was as high as 90 percent during the Eisenhower years and over 50 percent early in Ronald Reagan's administration, but fell to 35 percent in 2002, where it remained through 2012. While this shift obviously boosted the *after-tax* income of top earners, research suggests it can also affect reported *pre-tax* income by motivating people to work harder (because they keep more of each dollar they earn) and report more of their actual income, rather than seek ways to hide or shelter it (because the costs of reporting to tax authorities aren't as high as before).

Restrictions on trade have also decreased. Like cheaper telecommunications and transportation, this makes markets more global, allowing international superstars to more easily compete with, and drive out, local producers. When Kia poached Peter Schreyer from Audi in 2006, it was a signal that the market for talented automobile designers was increasingly global, not local.

Although the top 1 percent and 0.01 percent have seen record increases in their earnings, the superstar economy has faced a few headwinds. Perhaps the most important among these is the growth of the *long tail*—the increased availability of niche products and services. Technology has not just lowered marginal costs; in many cases it has also lowered fixed costs, inventory costs, and the costs of searching. Each of these changes makes it more attractive to offer a greater variety of products and services, filling small niches that previously went unfilled.

Instead of going head-to-head with a superstar, some individuals and businesses are instead finding ways to differentiate their products, to find or create an alternative niche where they can be the world's best. J. K. Rowling is a billion-dollar author, but there are also millions of other authors who now have a chance to publish for

---

rity, he is benefitting more directly from the technologies of superstardom discussed in the previous section.

more specialized audiences of a few thousand or even a few hundred readers. Amazon will stock their books and make them accessible to people across the planet. That will be profitable for Amazon even if it would have been unprofitable for any physical bookstore, with a much smaller set of customers, to stock the book. Even as the technology destroys geography—a barrier that used to protect authors from worldwide competition—it opens up specialization as a source of differentiation.

Instead of being the thousandth-best children's book author in the world, it may be more profitable to be the number-one author in Science-Based Advice for Ecological Entrepreneurs, or Football Clock Management.[20] Following this principle, developers have created over seven hundred thousand apps for the iPhone and Android, while Amazon offers over twenty-five million songs. An even larger number of blog posts, Facebook stories, and YouTube videos have been created in the sharing economy, creating economic value if not necessarily direct income for their creators. As we've seen, however, opportunities to create new products don't necessarily come with big paychecks. A superstar or long-tail economy with low barriers to entry is still one with far more inequality.

The Power Curve Nation

An economy dominated by winner-take-all markets has very different dynamics than the industrial economy to which we are accustomed. As we discussed at the beginning of the chapter, the earnings of bricklayers will vary a lot less than the winner-take-all earnings of app developers, but that's not the only difference. Instead of stable market shares, where revenues and income correspond proportionally to differences in talent and effort, competition in winner-take-all markets will be much more unstable and asymmetrical. The great economist Joseph Schumpeter wrote of "creative destruction," where each innovation not only created value for consumers but also wiped out the previous incumbent. The winners scaled up and dominated their markets, but were in turn vulnerable to the next generation of innovators. Schumpeter's observation describes markets in software, media, and the Internet much better than traditional markets in manufacturing and services. But as more and more industries become increasingly digitized and networked, we can expect the Schumpeterian dynamic to spread.[21]

In a superstar economy, the distribution of income isn't just more spread out; it has a very different shape. It's not just that a small group at the top sees big increases. It's also a change in the fundamental structure of the distribution. When revenues are roughly proportional to absolute performance, as in the example of the bricklayer, the earnings distribution is likely to roughly match the distribution of aptitude and effort. For many characteristics, humans fall roughly along a *normal distribution*, also known as the *Gaussian distribution* or the *bell curve*. That's the approximate distribution for height, strength, speed, general IQ, and in all likelihood many other characteristics such as emotional intelligence, management savvy, and even diligence.

Normal distributions are very common (hence the name), and they have an intuitive pattern. As you move further and further into either tail, the number of participants drops precipitously. What's more, the mean, median, and mode of the distribution are all the same number. An 'average' person is also the one in the middle of the

FIGURE 10.1

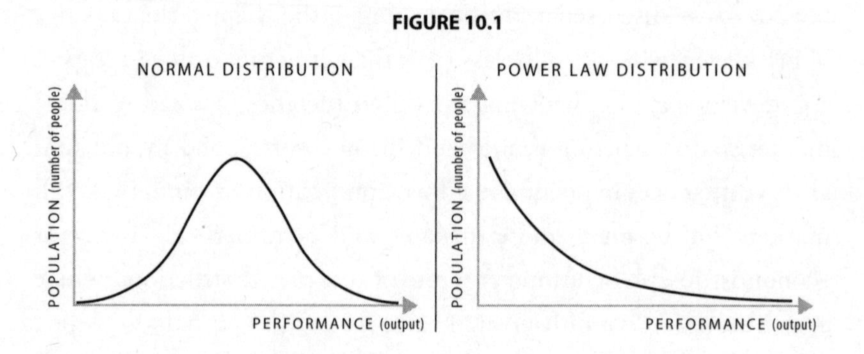

distribution, as well as the most typical or common type of person. If the income distribution of the United States followed a normal distribution, then median income would have risen along with average income—but of course, it didn't. Another characteristic of the normal distribution is that as you diverge from the mean, the probability of finding anyone with extreme characteristics drops rapidly, and at an increasing rate. The ratio of people who are seven feet tall to people who are six and a half feet tall is much less than the ratio people who are six and a half feet tall to people who are six feet tall. Thus, there are very few people at the extremes.

In contrast, superstar (and long tail) markets are often better described by a power law, or Pareto curve, in which a small number of people reap a disproportionate share of sales. This is often characterized as the 80/20 rule, where 20 percent of the participants get 80 percent of the gains, but it can be more extreme than that.[22] For instance, research by Erik and his coauthors found that book sales at Amazon were characterized by a power law distribution.[23] Power law distributions have a 'fat tail,' which means the likelihood of extreme events is much greater than one would expect to see in a normal distribution.[24] They are also 'scale invariant,' which means that the top-selling book accounts for about the same share of the top ten books' sales as the top ten books do for the top one hundred, or the top one hundred do for the top one thousand. Power laws describe many phenomena, from frequency of earthquakes to the frequency of words in most languages. They also describe the sales distribution of books, DVD, apps, and other information products.

Other markets are mixtures of different types of distributions. The U.S. economy as a whole can be described as a mixture of a log-normal distribution (a variant of the classical normal distribution) and power law, with the power law fitting the incomes at the top best.[25] Some of our current research at MIT is trying to better understand the causes and consequences of this mixture, and how it may be evolving over time.

A shift in the distribution of income to a power-law distribution would have important implications. For instance, Kim Taipale, founder of the Stilwell Center for Advanced Studies in Science and Technology Policy, has argued that, "The era of bell curve distributions that supported a bulging social middle class is over and we are headed for the power-law distribution of economic opportunities. Education per se is not going to make up the difference."[26]

Such a shift disrupts our mental models for understanding the world. Most of us are used to reasoning by reference to a prototypical. Politicians talk about the "average voter" and marketing managers talk about the "typical consumer." This works well for normal distributions where the most common value is near the average or, more formally, the mode and mean of the distribution are the same or nearly the same. However, the mean (or average) of a power-law distribution is generally much, much higher than the median or the mode.[27] For instance, in 2009, the average salary for major league baseball players was $3,240,206, roughly three times the median salary of $1,150,000.[28]

In practical terms, this means that when income is distributed according to a power law, most people will be below average—say goodbye, Lake Wobegon! Furthermore, over time, average income can increase without any increase in the median income or, for that matter, without any increase in income for most people. Power-law distributions don't just increase income inequality; they also mess with our intuitions.

CHAPTER 11

IMPLICATIONS OF THE BOUNTY AND THE SPREAD

"The test of our progress is not whether we add more to the abundance of those who have much it; is whether we provide enough for those who have little."

—Franklin D. Roosevelt

IN THE LAST FOUR chapters, we've seen that the second machine age contains a paradox. GDP has never been higher and innovation has never been faster, yet people are increasingly pessimistic about their children's future living standards. Adjusted for inflation, the combined net worth on *Forbes*' billionaire list has more than quintupled since 2000, but the income of the median household in America has fallen.[1]

The economic statistics underscore the dichotomy of bounty and spread. The economist Jared Bernstein, a senior fellow at the Center on Budget and Policy Priorities, brought our attention to the way productivity and employment have become decoupled, as shown in Figure 11.1. While these two key economic statistics tracked each other for most of the postwar period, they became decoupled in the late 1990s. Productivity continued its upward path as employment sagged. Today the employment-to-population ratio is lower than any time in at least 20 years, and the real income of the median worker is lower today than in the 1990s. Meanwhile, like productivity, GDP, corporate investment, and after-tax profits are also at record highs.

In a place like Silicon Valley or a research university like MIT, the rapid pace of innovation is particularly easy to see. Startups flourish, minting new millionaires and billionaires, while research labs churn out astonishing new technologies like the ones we saw in earlier chapters. At the same time, however, a growing number of peo-

FIGURE 11.1 Labor Productivity and Private Employment

ple face financial hardships: students struggle with enormous debt, recent graduates have difficulty finding new jobs, and millions have turned to debt to temporarily maintain their living standards.

In this chapter, we'll address three important questions about the future of the bounty and the spread. First, will the bounty overwhelm the spread? Second, can technology not only increase inequality but also create structural unemployment? And thirdly, what about globalization, the other great force transforming the economy—could it explain recent declines in wages and employment?

What's Bigger, Bounty or Spread?

Thanks to technology, we are creating a more abundant world—one where we get more and more output from fewer inputs like raw materials, capital, and labor. In the years to come we will continue to benefit in the form of things that are relatively easy to measure, such as higher productivity, and things that are less susceptible to metrics, such as the boost we get from free digital goods.

The previous paragraph describes our current bounty in the dry vocabulary of economics. This is a shame and needs to be corrected—a phenomenon so fundamental and wonderful deserves better language. 'Bounty' doesn't mean simply more cheap consumer goods and empty calories. As we noted in chapter 7, it also means simultaneously more choice, greater variety, and higher quality in many areas of our lives. It means heart surgeries performed without cracking the sternum and opening the chest cavity. It means constant access to the world's best teachers combined with personalized self-assessments that let students know how well they're mastering the material. It means that households have to spend less of their total budget over time on groceries, cars, clothing, and utilities. It means returning hearing to the deaf and, eventually, sight to the blind. It means less need to work doing boring, repetitive tasks and more opportunity for creative and interactive work.

The manifestations of progress are all based at least in part on digital technologies. When combined with political and economic systems that offer people choices instead of locking them in, technological advance is an awe-inspiring engine of betterment and bounty. It is also an engine driving spread, creating larger and larger differences over time in areas that we care about—wealth, income, standards of living, and opportunities for advancement. Some of these trends (particularly rising inequality) are also visible in other countries. We wish that progress in digital technologies were a rising tide that lifted all boats equally in all areas, but it's not.

Technology is certainly not the only force causing this rise in spread, but it is one of the main ones. Today's information technologies favor more-skilled over less-skilled workers, increase the returns to capital owners over labor, and increase the advantages that superstars have over everybody else. All of these trends increase spread—between those that have a job and those that don't, between highly skilled and educated workers and less advanced ones, between superstars and the rest of us. It's clear to us, from everything we've seen and learned recently, that all else being equal, future technologies will tend to increase spread, just as they will boost the bounty.

The fact that technology brings both bounty and spread, and brings more of both over time, leads to an important question: *Since there's so much bounty, should we be concerned about the spread?* In other words, we might consider rising inequality less of a problem if people at the bottom are also seeing their lives improve thanks to technology.

Income inequality and other measures of spread are increasing, but not everyone is convinced this is a problem. Some observers advance what we will call the 'strong bounty' argument, which essentially says that a focus on spread is misleading and inappropriate, since bounty is the more important phenomenon and exists even at the bottom of the spread. This argument acknowledges that highly skilled workers are pulling away from the rest—and that superstars are pulling so far away as to be out of sight—but then essentially asks, "So what? As long as all people's economic lives are getting better, why should we be concerned if some are getting a *lot* better?" As Harvard economist Greg Mankiw has argued, the enormous income earned by the "one percent" is not necessarily a problem if it reflects the just deserts of people who are creating value for everyone else.[2]

Capitalist economic systems work in part because they provide strong incentives to innovators: if your offering succeeds in the marketplace, you'll reap at least some of the financial rewards. And if your offering succeeds like crazy, the rewards can be huge. When these incentives are working well (and *not* doing things like providing huge, risk-free rewards to people taking inappropriate risks within the financial system), the benefits can be both large and broad: innovators improve the lives of many people whose purchases, in aggregate, make the innovator rich. Everyone benefits, even though not all benefits are the same.

The high-tech industry offers many examples of this happy phenomenon in action. Entrepreneurs create devices, websites, apps, and other goods and services that we value. We buy and use them in large numbers, and the entrepreneurs enjoy great financial success. This is not a dysfunctional pattern; it's a beneficial one. As economist Larry Summers put it, "suppose the United States had 30 more people like Steve Jobs—. . . . [W]e do need to recognize that a component of this inequality is the other side of successful entrepreneurship; that is surely something we want to encourage."[3]

We particularly want to encourage it because, as we saw in chapter 6, technological progress typically helps even the poorest people around the world. Careful research has shown that innovations like mobile telephones are improving people's incomes, health, and other measures of well-being. As Moore's Law continues to simultaneously drive down the cost and increase the capability of these devices, the benefits they bring will continue to add up.

If the strong bounty argument is correct, then we have nothing significant to worry about as we head deeper into the second machine age. But is it? We wish that were the case, but it's not. As we saw in chapters 9 and 10, the data are quite clear that many people in the United States and elsewhere are losing ground over time, not just relative to others but in absolute terms. In America, the income of the median worker is lower in real dollars than it was in 1999 and the story largely repeats itself when we look at households instead of individual workers, or total wealth instead of annual income. Many people are falling behind as technology races ahead.

Some proponents of the strong bounty argument believe that while these declines are real, they're still less important than the unmeasured price decreases, quality improvements, and other benefits that we've been experiencing. Economists Donald Boudreaux and Mark Perry write that:

> Spending by households on many of modern life's "basics"—food at home, automobiles, clothing and footwear, household furnishings and equipment, and housing and utilities—fell

> from 53% of disposable income in 1950 to 44% in 1970 to 32% today. . . [and] the quantities and qualities of what ordinary Americans consume are closer to that of rich Americans than they were in decades past. Consider the electronic products that every middle-class teenager can now afford—iPhones, iPads, iPods and laptop computers. They aren't much inferior to the electronic gadgets now used by the top 1% of American income earners, and often they are exactly the same.[4]

Perry adds that "thanks to innovation and technology . . . all Americans (especially low-income and middle-income groups) are better off today than in any previous period."[5] In the *National Review* and elsewhere, Scott Winship of the Brookings Institution has made similar points.[6]

These are intriguing arguments. We particularly like the insight that the average worker today is in important ways better off than his or her counterpart in earlier generations precisely because of the bounty brought by innovation and technology. For anything related to information, media, communication, and computation, the improvements are so large that they can hardly be believed in retrospect, or anticipated in advance. And the bounty doesn't stop there: technological progress also causes cost and quality improvements in other areas, such as food and utilities, that may not seem high-tech on the surface but actually are when you look under the hood.

These points have merit, but we are not convinced that people at the lower ranges of the spread are doing OK. For one thing, some critical items that they (and everyone else) would like to purchase are getting much more expensive over time. This phenomenon is well summarized in research by Jared Bernstein, who compared increases in median family income between 1990 and 2008 with changes in the cost of housing, health care, and college. He found that while family income grew by around 20 percent during that time, prices for housing and college grew by about 50 percent, and health care by more than 150 percent.[7] Since American real median incomes have been falling in recent years, these comparisons would be even more unfavorable if repeated over later time periods than 1990 to 2008.

However American households are spending their money, many of them are left without a financial cushion. The economists Annamaria Lusardi, Daniel J. Schneider, and Peter Tufano conducted a 2011 study asking people about "their capacity to come up with $2,000 in 30 days." Their findings are troubling. They concluded that, "Approximately one quarter of Americans report that they would certainly not be able to come up with such funds, and an additional 19% would do so by relying at least in part on pawning or selling possessions or taking payday loans. . . . [In other words, we] find that nearly half of Americans are financially fragile. . . . [A] sizable fraction of seemingly 'middle class' Americans . . . judge themselves to be financially fragile."[8]

Other data—about poverty rates, access to health care, the number of people who want full-time jobs but can only find part-time work, and so on—confirm the impression that while the economic bounty from technology is real, it is not sufficient to compensate for huge increases in spread. And those increases are not purely a consequence of the Great Recession, nor a recent or transient phenomenon.

That many Americans face stagnant and falling incomes is bad enough, but it is now combined with decreasing social mobility—an ever lower chance that children born at the bottom end of the spread will escape their circumstances and move upward throughout their lives and careers. Recent research makes it clear that the American Dream of upward mobility, which was real in earlier generations, is greatly diminished today. To take just one example, a 2013 study of U.S. tax returns from 1987 to 2009 conducted by economists Jason DeBacker, Bradley Heim, and their colleagues found that the thirty-five thousand households they studied tended to stay in roughly the same order of richest to poorest year after year, with little reshuffling, even as the differences in household income grew over time.[9] More recently, sociologist Robert Putnam has illustrated how for Americans in cities like Port Clinton, Ohio (his hometown), economic conditions and prospects have worsened in recent decades for the children of parents with only high school educations even as they've improved for college-educated families. This is exactly what we'd expect to see as skill-biased technical change accelerates.[10]

Many Americans believe that they still live in the land of opportunity—the country that offers the greatest chance of economic advancement. But this is no longer the case. As *The Economist* sums it up, "Back in its Horatio Alger days, America was more fluid than Europe. Now it is not. Using one-generation measures of social mobility—how much a father's relative income influences that of his adult son—America does half as well as Nordic countries, and about the same as Britain and Italy, Europe's least-mobile places."[11] So the spread is not only large, but also self-perpetuating. Too often, people at the bottom and middle stay where they are over their careers, and families stay locked in across generations. This is not healthy for an economy or society.

It would be even unhealthier if the spread were to diminish the bounty—if inequality and its consequences somehow impeded technological progress, keeping us from enjoying all the potential benefits of the new machine age. Although a common argument is that high levels of inequality can motivate people to work harder, boosting overall economic growth, the inequality can also dampen growth. In 2012 economist Daron Acemoglu and political scientist James Robinson published *Why Nations Fail*, a sweeping account of hundreds of years of history aimed at uncovering, as the book's subtitle puts it, "the origins of power, prosperity, and poverty." According to Acemoglu and Robinson, the true origins are not geography, natural resources, or culture. Instead, they're institutions like democracy, property rights, and the rule of law; inclusive ones bring prosperity, and extractive ones—ones that bend the economy and the rules of the game to the service of entrenched elite—bring poverty. The authors make a compelling case, and when they turn their attention to America's current condition, they offer important insights and cautions:

> Prosperity depends on innovation, and we waste our innovative potential if we do not provide a level playing field for all: we don't know where the next Microsoft, Google, or Facebook will come from, and if the person who will make this happen goes to a failing school and cannot get into a good university, the chances that it will become a reality are much diminished. . . .
>
> The U.S. generated so much innovation and economic growth for the last two hundred years because, by and large, it rewarded innovation and investment. This did not happen in a vacuum; it was supported by a particular set of political arrangements—inclusive political institutions—which prevented an elite or another narrow group from monopolizing political power and using it for their own benefit and at the expense of society.
>
> So here is the concern: economic inequality will lead to greater political inequality, and those who are further empowered politically will use this to gain greater economic advantage, stacking the cards in their favor and increasing economic inequality still further—a quintessential vicious circle. And we may be in the midst of it.[12]

Their analysis hits on a final reason to be concerned about the large and growing inequality of recent years: it could lead to the creation of extractive institutions that would slow our journey into the second machine age. We think this would be something more than a shame; it would be closer to a tragedy. We also believe, based on the work of Acemoglu and Robinson and others, that it is a plausible scenario. Instead of being confident that the bounty from technology will more than compensate for the spread it generates, we are instead concerned about something close to the reverse: that the spread could actually reduce the bounty in years to come.

Technological Unemployment

We've seen that the overall pie of the economy is growing, but some people, even a majority of them, can be made worse off by advances in technology. As demand falls for labor, particularly relatively unskilled labor, wages fall. But can technology actually lead to unemployment?

We're not the first people to ask these questions. In fact, they've been debated vigorously, even violently, for at least two hundred years. Between 1811 and 1817, a group of English textile workers whose jobs were threatened by the automated looms of the first Industrial Revolution rallied around a perhaps mythical, Robin Hood–like figure named Ned Ludd and attacked mills and machinery before being suppressed by the British government.

Economists and other scholars saw in the Luddite movement an early example of a broad and important new pattern: large-scale automation entering the workplace and affecting people's wage and employment prospects. Researchers soon fell into two camps. The first and largest argued that while technological progress and other factors definitely cause some workers to lose their jobs, the fundamentally creative nature of capitalism creates other, usually better, opportunities for them. Unemployment, therefore, is only temporary and not a serious problem. John Bates Clark (after whom the medal for the best economist under the age of forty is named) wrote in 1915 that "In the actual [economy], which is highly dynamic, such a supply of unemployed labor is always at hand, and it is neither possible [nor] normal that it should be altogether absent. The well-being of workers requires that progress should go on, and it cannot do so without causing temporary displacement of laborers."[13]

The following year, the political scientist William Leiserson took this argument further. He described unemployment as something close to a mirage: "the army of the unemployed is no more unemployed than are firemen who wait in fire-houses for the alarm to sound, or the reserve police force ready to meet the next call."[14] The creative forces of capitalism, in short, required a supply of ready labor, which came from people displaced by previous instances of technological progress.

John Maynard Keynes was less confident that things would always work out so well for workers. His 1930 essay "Economic Possibilities for our Grandchildren," while mostly optimistic, nicely articulated the position of the second camp—that automation could in fact put people out of work permanently, especially if more and more things kept getting automated. His essay looked past the immediate hard times of the Great Depression and offered a prediction: "We are being afflicted with a new disease of which some readers may not yet have heard the name, but of which they will hear a great deal in the years to come—namely, *technological unemployment*. This means unemployment due to our discovery of means of economizing the use of labor outrunning the pace at which we can find new uses for labor."[15] The extended joblessness of the Great Depression seemed to confirm Keynes's ideas, but it eventually eased. Then came World War II and its insatiable demands for labor, both on the battlefield and the home front, and the threat of technological unemployment receded.

After the war ended, the debate about technology's impact on the labor force resumed and took on new life once computers appeared. A commission of scientists and social theorists sent an open letter to President Lyndon Johnson in 1964 arguing that:

> A new era of production has begun. Its principles of organization are as different from those of the industrial era as those of the industrial era were different from the agricultural. The cybernation revolution has been brought about by the com-

> bination of the computer and the automated self-regulating machine. This results in a system of almost unlimited productive capacity which requires progressively less human labor.[16]

The Nobel Prize–winning economist Wassily Leontief agreed, stating definitively in 1983 that "the role of humans as the most important factor of production is bound to diminish in the same way that the role of horses in agricultural production was first diminished and then eliminated by the introduction of tractors."[17]

Just four years later, however, a panel of economists assembled by the National Academy of Sciences disagreed with Leontief and made a clear, comprehensive, and optimistic statement in their report "Technology and Employment":

> By reducing the costs of production and thereby lowering the price of a particular good in a competitive market, technological change frequently leads to increases in output demand: greater output demand results in increased production, which requires more labor, offsetting the employment effects of reductions in labor requirements per unit of output stemming from technological change. . . . Historically and, we believe, for the foreseeable future, reductions in labor requirements per unit of output resulting from new process technologies have been and will continue to be outweighed by the beneficial employment effects of the expansion in total output that generally occurs.[18]

This view—that automation and other forms of technological progress in aggregate create more jobs than they destroy—has come to dominate the discipline of economics. To believe otherwise is to succumb to the "Luddite Fallacy." So in recent years, most of the people arguing that technology is a net job destroyer have not been mainstream economists.

The argument that technology cannot create ongoing structural unemployment, rather than just temporary spells of joblessness during recessions, rests on two pillars: 1) economic theory and 2) two hundred years of historical evidence. But both of these are less solid than they first appear.

First, the theory. There are three economic mechanisms that are candidates for explaining technological unemployment: inelastic demand, rapid change, and severe inequality.

If technology leads to more efficient use of labor, then as the economists on the National Academy of Sciences panel pointed out, this does not automatically lead to reduced demand for labor. Lower costs may lead to lower prices for goods, and in turn, lower prices lead to greater demand for the goods, which can ultimately lead to an increase in demand for labor as well. Whether or not this will actually happen depends on the elasticity of demand, defined as the percentage increase in the quantity demanded for each percentage decline in price.

For some goods and services, such as automobile tires or household lighting, demand has been relatively inelastic and thus insensitive to price declines.[19] Cutting the price of artificial light in half did not double the amount of light consumers and businesses demanded, so the total revenues for the lighting industry have fallen as lighting became more efficient. In an great piece of historical sleuthing, economist William Nordhaus documented how technology has reduced the price of light by over a thousand-fold since the days of candles and whale oil lamps, allowing us to expend far less on labor while getting all the light we need.[20] Whole sectors of the economy, not just product categories, can face relatively inelastic demand. Over the years agriculture and manufacturing have each experienced falling employment as they became more efficient. The lower prices and improved quality of their outputs did not lead to enough increased demand to offset improvements in productivity.

On the other hand, when demand is very elastic, greater productivity leads to enough of an increase in demand that more labor ends up employed. The possibility of this happening for some types of energy has been called the Jevons paradox: more energy efficiency can sometimes lead to greater total energy consumption. But to economists there is no paradox, just an inevitable implication of elastic demand. This is especially common in new industries like information technology.[21] If elasticity is exactly equal to one (i.e., a 1 percent decline in price leads to exactly a 1 percent increase in quantity), then total revenues (price times quantity) will be unchanged. In other words, an increase in productivity will be exactly matched by an identical increase in demand to keep everyone just as busy as they were before.

Elasticity of exactly one might seem like a very special case, yet a good (but not airtight) argument can be made that, in the long run, this is exactly what happens in the overall economy. For instance, falling food prices might reduce demand for agricultural labor, but they free up just enough money to be spent elsewhere in the economy so that overall employment is maintained.[22] The money is spent not just buying more of the existing goods, but also on newly invented products and services. This is the core of the economic argument that technological unemployment is impossible.

KEYNES DISAGREED. He thought that in the long run, demand would not be perfectly inelastic. That is, ever lower (quality-adjusted) prices would not necessarily mean we would consume ever more goods and services. Instead, we would become satiated and choose to consume less. He predicted that this would lead to a dramatic reduction in working hours to as few as fifteen per week as less and less labor was needed to produce all the goods and services that people demanded.[23] However, it's hard to see this type of technological unemployment as an economic problem. After all, in that scenario, by definition, people are working less because they are satiated. The "economic problem" of scarcity is replaced by the entirely more appealing problem of what to do with abundant wealth and copious leisure. As Arthur C. Clarke is purported to have put it, "The goal of the future is full unemployment, so we can play."[24]

Keynes was more concerned with short-term "maladjustments," which brings us to the second, more serious argument for technological unemployment: the inability of our skills, organizations, and institutions to keep pace with technical change. When technology eliminates one type of job, or even the need for a whole category of skills, those workers will have to develop new skills and find new jobs. Of course, that can take time, and in the meantime they may be unemployed. The optimistic argument maintains that this is temporary. Eventually, the economy will find a new equilibrium and full employment will be restored as entrepreneurs invent new businesses and the workforce adapts its human capital.

But what if this process takes a decade?[25] And what if, by then, technology has changed again? This is the possibility that Wassily Leontief had in mind his 1983 article when he speculated that many workers could end up permanently unemployed, like horses unable to adjust to the invention of the tractors.[26] Once one concedes that it takes time for workers and organizations to adjust to technical change, then it becomes apparent that accelerating technical change can lead to widening gaps and increasing possibilities for technological unemployment. Faster technological progress may ultimately bring greater wealth and longer lifespans, but it also requires faster adjustments by both people and institutions. With apologies to Keynes, in the long run we may not be dead, but we will still need jobs.

The third argument for technological unemployment may be the most troubling of all. It goes beyond "temporary" maladjustments. As described in detail in chapters 8 and 9, recent advances in technology have created both winners and losers via skill-biased technical change, capital-biased technical change, and the proliferation of superstars in winner-take-all markets. This has reduced the demand for some types of work and skills. In a free market, prices adjust to restore equilibrium between supply and demand, and indeed, real wages have fallen for millions of people in the United States.

In principle, the equilibrium wage could be one dollar an hour for some workers, even as other workers command a wage thousands of times higher. Most people in advanced countries would not consider one dollar an hour a living wage, and don't expect society to require people to work at that wage under threat of starvation. What's more, in extreme winner-take-all markets, the equilibrium wage might be zero: even if we offered to sing "Satisfaction" for free, people would still prefer to pay for the version sung by Mick Jagger. In the market for music, Mick can now, in effect, make digital copies of himself that compete with us. A near-zero wage is not a living wage. Rational people would rather look for another gig, and look, and look, and look, than depend on a near-zero wage for their sustenance.

Thus, there is a floor on how low wages for human labor can go. In turn, that floor can lead to unemployment: people who want to work, but are unable to find jobs. If neither the worker nor any entrepreneur can think of a profitable task that requires that worker's skills and capabilities, then that worker will go unemployed indefinitely. Over history, this has happened to many other inputs to production that were once valuable, from whale oil to horse labor. They are no longer needed in today's economy even at zero price. In other words, just as technology can create inequality, it can also create unemployment. And in theory, this can affect a large number of people, even a majority of the population, and even if the overall economic pie is growing.

So that's theory, but what about the data? For most of the two hundred years since the Luddite rebellion technology has boosted productivity enormously, but the data show that employment grew alongside productivity up until the end of the twentieth century. This shows that productivity doesn't always lead to job destruction. It's even tempting to suppose that productivity somehow inevitably leads to job creation, as technology boosters sometimes argue. However, as we saw in figure 11.1, the data also show that, more recently, job growth decoupled from productivity in the late 1990s. According to Jared Bernstein, the anti-Luddites call this fact a "head scratcher." Which history should we take guidance from: the two centuries ending in the late 1990s, or the fifteen years since then? We can't know for sure, but our reading of technology tells us that the power of exponential, digital, and combinatorial forces, as well as the dawning of machine intelligence and networked intelligence, presage even greater disruptions.

The Android Experiment

Imagine that tomorrow a company introduced androids that could do absolutely everything a human worker could do, including building more androids. There's an endless supply of these robots, and they're extremely cheap to buy and virtually free to run over time. They work all day, every day, without breaking down.

Clearly, the economic implications of such an advance would be profound. First of all, productivity and output would skyrocket. The androids would operate the farms and factories. Food and products would become much cheaper to produce. In a competitive market, in fact, their prices would fall close to the cost of their raw materials. Around the world, we'd see an amazing increase in the volume, variety, and affordability of offerings. The androids, in short, would bring great bounty.

They'd also bring severe dislocations to the labor force. Every economically rational employer would prefer androids, since compared to the status quo they provide equal capability at lower cost. So they would very quickly replace most, if not all, human workers. Entrepreneurs would continue to develop novel products, create new markets, and found companies, but they'd staff these companies with androids instead of people. The owners of the androids and other capital assets or natural resources would capture all the value in the economy, and do all the consuming. Those with no assets would have only their labor to sell, and their labor would be worthless.

This thought experiment reflects the reality that there is no 'iron law' that technological progress must always be accompanied by broad job creation.

One slight variation on this thought experiment imagines that the androids can do everything a human worker can do except for one skill—say, cooking. The economic results would be unchanged, except that there would still be human cooks. Because there would be so much competition for these jobs, however, companies that employed cooks could offer much lower wages and still fill their open positions. The total number of hours spent cooking in the economy would stay the same (at least as long as people kept eating in restaurants), but the total wages paid to cooks would go down. The only exception might be superstar chefs with some combination of skill and reputation that could not be duplicated by other people. Superstars would still be able to command high wages; other cooks would not. So in addition to bringing great bounty of output, the androids would also greatly increase the spread in income.

How useful are these thought experiments, which sound more like science fiction than any current reality? Fully functional humanoid androids are not rumbling around at American companies today. In fact, they don't yet exist, and until recently progress had been slow in making machines that can take the places of human workers in areas like pattern recognition, complex communication, sensing, and mobility. But as we've seen, the pace of progress here has been accelerating greatly in recent years.

The better machines can substitute for human workers, the more likely it is that they'll drive down the wages of humans with similar skills. The lesson from economics and business strategy is that you don't want to compete against close substitutes, especially if they have a cost advantage.

But in principle, machines can have very different strengths and weaknesses than humans. When engineers work to amplify these differences, building on the areas where machines are strong and humans are weak, then the machines are more likely to *complement* humans rather than substitute for them. Effective production is more likely to require both human and machine inputs, and the value of the human inputs will grow, not shrink, as the power of machines increases. A second lesson of economics and business strategy is that it's great to be a complement to something that's increasingly plentiful. Moreover, this approach is more likely to create opportunities to produce goods and services that could never have been created by unaugmented humans, or machines that simply mimicked people, for that matter. These new goods and services provide a path for productivity growth based on increased output rather than reduced inputs.

Thus in a very real sense, as long as there are unmet needs and wants in the world, unemployment is a loud warning that we simply aren't thinking hard enough about what needs doing. We aren't being creative enough about solving the problems we have using the freed-up time and energy of the people whose old jobs were automated away. We can do more to invent technologies and business models that augment and amplify the unique capabilities of humans to create new sources of value, instead of automating the ones that already exist. As we will discuss further in the next chapters, this is the real challenge facing our policy makers, our entrepreneurs, and each of us individually.

An Alternative Explanation: Globalization

Technology isn't the only thing transforming the economy. The other big force of our era is globalization. Could this be the reason that median wages have stagnated in the United States and other advanced economies? A number of thoughtful economists have made exactly that argument. The story is one of *factor price equalization*.

This means that in any single market, competition will tend to bid the prices of the factors of production—such as labor or capital—to a single, common price.* Over the past few decades, lower transaction in communication costs have helped create one big global market for many products and services.

Businesses can identify and hire workers with skills they need anywhere in the world. If a worker in China can do the same work as an American, then what economists call "the law of one price" demands that they earn essentially the same wages, because the market will arbitrage away differences just as it would for other commodities. That's good news for the Chinese worker, and for overall economic efficiency. But is not good news for the American worker who now faces low-cost competition. A number of economists have made exactly this argument. Michael Spence, in his brilliant book *The Next Convergence*, explains how the integration of global markets is leading to enormous dislocations, especially in labor markets.[27]

The factor price equalization story yields a testable prediction: American manufacturers would be expected to shift production overseas, where costs are lower. And indeed manufacturing employment in the United States has fallen over the past twenty years; economists David Autor, David Dorn, and Gordon Hanson estimate that competition from China can explain about a quarter of the decline in U.S. manufacturing employment.[28] However, when one looks more closely at the data, the globalization story becomes much less compelling. Since 1996, manufacturing employment in China itself has actually fallen as well, coincidentally by an estimated 25 percent.[29] That's over thirty million fewer Chinese workers in that sector, even while output soared by 70 percent. It's not that American workers are being replaced by Chinese workers. It's that both American and

* This is no different from the concept we invoked when we were comparing and equating the wages of human workers with robots that, hypothetically, had identical capabilities.

Chinese workers are being made more efficient by automation. As a result, both countries are producing more output with fewer workers.

In the long run, the biggest effect of automation is likely to be on workers not in America and other developed nations, but rather in developing nations that currently rely on low-cost labor for their competitive advantage. If you take most of the costs of labor out of the equation by installing robots and other types of automation, then the competitive advantage of low wages largely disappears. This is already beginning to happen. Terry Guo of Foxconn has been aggressively installing hundreds of thousands of robots to replace an equivalent number of human workers. He says he plans to buy millions more robots in the coming years. The first wave is going into factories in China and Taiwan, but once an industry becomes largely automated, the case for locating a factory in a low-wage country becomes less compelling. There may still be logistical advantages if the local business ecosystem is strong, making it easier to get spare parts, supplies, and custom components. But over time inertia may be overcome by the advantages of reducing transit times for finished products and being closer to customers, engineers and designers, educated workers, or even regions where the rule of law is strong. This can bring manufacturing back to America, as entrepreneurs like Rod Brooks have been emphasizing.

A similar argument applies outside of manufacturing. For instance, interactive voice-response systems are automating jobs in call centers. United Airlines has been successful in making such a transition. This can disproportionally affect low-cost workers in places like India and the Philippines. Similarly, many medical doctors used to have their dictation sent overseas to be transcribed. But an increasing number are now happy with computer transcription. In more and more domains, intelligent and flexible machines, not humans in other countries, are the most cost-effective source for 'labor.'

If you look at the types of tasks that have been offshored in the past twenty years, you see that they tend to be relatively routine, well-structured tasks. Interestingly, these are precisely the tasks that are easiest to automate. If you can give precise instructions to someone else on exactly what needs to be done, you can often write a precise computer program to do the same task. In other words, offshoring is often only a way station on the road to automation.

In the long run, low wages will be no match for Moore's Law. Trying to fend off advances in technology by cutting wages is only a temporary protection. It is no more sustainable than asking folk legend John Henry to lift weights to better compete with a steam-powered hammer.

CHAPTER 12

LEARNING TO RACE *WITH* MACHINES: RECOMMENDATIONS FOR INDIVIDUALS

"But they are useless. They can only give you answers."

—Pablo Picasso, on computers[1]

We've talked about our research findings and conclusions with many different groups, from executive teams to radio show audiences. Almost every time we do, one of the first questions is something like, "I have children in school. How should I be helping them prepare for the future you're describing?" Sometimes the kids are in college, sometimes they're in kindergarten, but the question is the same. And it's not just parents who are concerned about career opportunities in the second machine age. Students themselves, leaders of the organizations that might hire them, educators, policy makers and elected officials, and many others also wonder which human skills and abilities, if any, will still be valued as technology continues to improve.

Recent history shows that this is a difficult question to answer. Frank Levy and Richard Murnane's excellent book *The New Division of Labor* was by far the best research and thinking on this topic when it came out in 2004, arguing that pattern recognition and complex communication were the two broad areas where humans would continue to hold the high ground over digital labor. As we've seen, however, this has not always proved to be the case. So as technology races ahead, will it leave a generation behind in all areas, or at least most of them?

The answer is no. Even in those areas where digital machines have far outstripped humans, people still have vital roles to play. This sounds like a contradiction in terms; the game of chess shows why it's not.

Even Though It's Checkmate, It's Not Game Over

After the reigning world champion Garry Kasparov lost to the IBM computer Deep Blue in 1997, head-to-head contests between people and chess computers lost much of their allure; it was clear that future competitions would be increasingly one-sided. Dutch grandmaster Jan Hein Donner summed up the current attitude of human chess masters. When asked how he would prepare for a match against a computer, he replied, "I would bring a hammer."[2]

It might seem, then, that humans no longer have anything to contribute to the game of chess. But the invention of 'freestyle' chess tournaments shows how far this is from the truth. In these events, teams can include any combination of human and digital players. As Kasparov himself explains when discussing the results of a 2005 freestyle contest,

> The teams of human plus machine dominated even the strongest computers. The chess machine Hydra, which is a chess-specific supercomputer like Deep Blue, was no match for a strong human player using a relatively weak laptop. Human strategic guidance combined with the tactical acuity of a computer was overwhelming.
>
> The surprise came at the conclusion of the event. The winner was revealed to be not a grandmaster with a state-of-the-art PC but a pair of amateur American chess players using three computers at the same time. Their skill at manipulating and "coaching" their computers to look very deeply into positions effectively counteracted the superior chess understanding of their grandmaster opponents and the greater computational power of other participants. Weak human + machine + better

> process was superior to a strong computer alone and, more remarkably, superior to a strong human + machine + inferior process.[3]

The key insight from freestyle chess is that people and computers don't approach the same task the same way. If they did, humans would have had nothing to add after Deep Blue beat Kasparov; the machine, having learned how to mimic human chess-playing ability, would just keep riding Moore's Law and racing ahead. But instead we see that people still have a great deal to offer the game of chess at its highest levels once they're allowed to race with machines, instead of purely against them.

So what are these still-valuable, uniquely human abilities? Kasparov writes about human "strategic guidance" vs. computers' "tactical acuity" in chess, but the distinction between these two is often not clear, particularly in advance. Similarly, as we noted earlier, technology has made deeper inroads into routine tasks than nonroutine work.

This distinction is a valid and important one—adding up a column of numbers is totally routine and by now totally automated—but here again the boundary between the two task categories is not always obvious. Very few people, for example, would have considered playing chess a 'routine' task half a century ago. In fact, it was considered one of the highest expressions of human ability. As the former world champion Anatoly Karpov wrote about the idols of his youth, "I simply lived in one world, and the grandmasters existed in a completely different one. People like that were not really even people, but like gods or mythical heroes."[4] But the human heroes fell to routine, number-crunching computers in this domain. And yet, once they were allowed to work with machines instead of only against them, they reasserted their value. How?

Eureka—Something Computers Can't Do!

Kasparov offers an important clue when describing a match he played against the Bulgarian grandmaster Veselin Topalov, during which they were each allowed to freely consult a computer. Kasparov knew, he wrote, that "since we both had equal access to the same database, the advantage still came down to creating a new idea at some point."[5] As we look across examples of things we haven't seen computers do yet, this idea of the "new idea" keeps recurring.

We've never seen a truly creative machine, or an entrepreneurial one, or an innovative one. We've seen software that could create lines of English text that rhymed, but none that could write a true poem ("the spontaneous overflow of powerful feelings, recollected in tranquility," as Wordsworth described it). Programs that can write clean prose are amazing achievements, but we've not yet seen one that can figure out what to write about next. We've also never seen software that could create good software; so far, attempts at this have been abject failures.

These activities have one thing in common: *ideation*, or coming up with new ideas or concepts. To be more precise, we should probably say *good* new ideas or concepts, since computers can easily be programmed to generate new combinations of preexisting elements like words. This however, is not recombinant innovation in any meaningful sense. It's closer to the digital equivalent of a hypothetical room full of monkeys banging away randomly on typewriters for a million years and still not reproducing a single play of Shakespeare's.

Ideation in its many forms is an area today where humans have a comparative advantage over machines. Scientists come up with new hypotheses. Journalists sniff out a good story. Chefs add a new dish to the menu. Engineers on a factory floor figure out why a machine is no longer working properly. Steve Jobs and his colleagues at Apple figure out what kind of tablet computer we actually want. Many of these activities are supported and accelerated by computers, but none are driven by them.

Picasso's quote at the head of this chapter is just about half right. Computers are not useless, but they're still machines for generating answers, not posing interesting new questions. That ability still seems to be uniquely human, and still highly valuable. We predict that people who are good at idea creation will continue to have a comparative advantage over digital labor for some time to come, and will find themselves in demand. In other words, we believe that employers now and for some time to come will, when looking for talent, follow the advice attributed to the Enlightenment sage Voltaire: "Judge a man by his questions, not his answers."[6]

Ideation, creativity, and innovation are often described as 'thinking outside the box,' and this characterization indicates another large and reasonably sustainable advantage of human over digital labor. Computers and robots remain lousy at doing anything outside the frame of their programming. Watson, for example, is an amazing *Jeopardy!* player, but would be defeated by a child at *Wheel of Fortune*, *The Price is Right*, or any other TV game show unless it was substantially reprogrammed by its human creators. Watson is not going to get there on its own.

Instead of conquering other game shows, however, the IBM team behind Watson is turning its attention to other fields such as medicine. Here again, it will be limited by its frame. Make no mistake: we believe that Watson will ultimately make an excellent doctor. Right now human diagnosticians reign supreme, but just as Watson soon got good enough to beat Ken Jennings, Brad Rutter, and all other human *Jeopardy!* players, we predict that Dr. Watson will soon be able to beat Dr. Welby, Dr. House, and real human doctors at their own game.

While computer reasoning from predefined rules and inferences from existing examples can address a large share of cases, human diagnosticians will still be valuable even after Dr. Watson finishes its medical training because of the idiosyncrasies and special cases that inevitably arise. Just as it is much harder to create a 100-percent self-driving car than one that merely drives in normal conditions on a highway, creating a machine-based system for covering all possible medical cases is radically more difficult than building one for the most common situations. As with chess, a partnership between Dr. Watson and a human doctor will be far more creative and robust than either of them working alone. As futurist Kevin Kelly put it "You'll be paid in the future based on how well you work with robots."[7]

Sensing Our Advantage

So computers are extraordinarily good at pattern recognition within their frames, and terrible outside them. This is good news for human workers because thanks to our multiple senses, our frames are inherently broader than those of digital technologies. Computer vision, hearing, and even touch are getting exponentially better all the time, but there are still tasks where our eyes, ears, and skin, to say nothing of our noses and tongues, surpass their digital equivalents. At present and for some time to come, the sensory package and its tight connection to the pattern-recognition engine of the brain gives us a broader frame.

The Spanish clothing company Zara exploits this advantage and uses humans instead of computers to decide which clothes to make. For most apparel retailers, forecasting and sales planning are largely statistical affairs, conducted months in advance of the clothes actually showing up in stores. Zara takes a different approach. It specializes in 'fast fashion'—inexpensive, trendy clothes aimed primarily at teens and young adults. Because these styles gain popularity as quickly as they fade away, Zara has configured its factories and warehouses to make and deliver garments very rapidly, while they're still hot. To answer the critical question "Which clothes should we make and ship to each store?" Zara relies on its store managers around the world to order exactly, and only, the merchandise that will sell in that location over the next few days.[8]

Managers figure this out not by consulting algorithms but instead by walking around the store, observing what shoppers (particularly cool ones) are wearing, talking to them about what they like and what they're looking for, and generally doing many things at which people excel. Zara store managers do a lot of visual pattern recognition, engage in complex communication with customers, and use all of this information for two purposes: to order existing clothes using a broad frame of inputs, and to engage in ideation by telling headquarters what kinds of new clothes would be popular in their location. Zara has no plans to switch from human-based to machine-based ordering any time soon, and we think they're making a very smart decision.

So ideation, large-frame pattern recognition, and the most complex forms of communication are cognitive areas where people still seem to have the advantage, and also seem likely to hold on to it for some time to come. Unfortunately, though, these skills are not emphasized in most educational environments today. Instead, primary education often focuses on rote memorization of facts, and on the skills of reading, writing, and arithmetic—the 'three *Rs*,' as Tory MP Sir William Curtis named them around 1825 (incidentally, it's unlikely that a machine would have given them a moniker as memorable, if technically inaccurate, as the 'three *Rs*').[9]

To Switch the Skills, Switch the Schools

Education researcher Sugata Mitra, who has showed how much poor children in the developing world can learn on their own when provided with nothing more than some appropriate technology, has a provocative explanation for the emphasis on rote learning. In his speech at the 2013 TED conference, where his work was recognized with the one-million-dollar TED prize, he gave an account of when and why these skills came to be valued.

> I tried to look at where did the kind of learning we do in schools, where did it come from? . . . It came from . . . the last and the biggest of the empires on this planet, [the British Empire].
>
> What they did was amazing. They created a global computer made up of people. It's still with us today. It's called the bureaucratic administrative machine. In order to have that machine running, you need lots and lots of people. They made another machine to produce those people: the school. The schools would produce the people who would then become parts of the bureaucratic administrative machine. . . . They must know three things: They must have good handwriting, because the data is handwritten; they must be able to read; and they must be able to do multiplication, division, addition and subtraction in their head. They must be so identical that you could pick one up from New Zealand and ship them to Canada and he would be instantly functional.[10]

Of course, we like this explanation because it describes things as computers and machines. But more fundamentally, we like it because it points out that the three *R*s were once the skills that workers needed to contribute to the most advanced economy of the time. As Mitra points out, the educational system of Victorian England was designed quite well for its time and place. But that time and place are no longer ours. As Mitra continued:

> The Victorians were great engineers. They engineered a system that was so robust that it's still with us today, continuously producing identical people for a machine that no longer exists. . . . [Today] the clerks are the computers. They're there in thousands in every office. And you have people who guide those computers to do their clerical jobs. Those people don't need to be able to write beautifully by hand. They don't

> need to be able to multiply numbers in their heads. They do need to be able to read. In fact, they need to be able to read discerningly.[11]

Mitra's work shows that children, even poor and uneducated ones, can learn to read discerningly. The children in his studies form teams, use technology to search broadly for relevant information, discuss what they're learning with one another, and eventually come up with new (to them) ideas that very often turn out to be correct. In other words, they acquire and demonstrate the skills of ideation, broad-frame pattern recognition, and complex communication. So the "self-organizing learning environments" (SOLEs) Mitra observed seem to be teaching children the skills that will give them advantages over digital labor.

We probably shouldn't be too surprised by this; SOLEs have been around for a while, and have produced many people who have excelled at racing with machines. In the early years of the twentieth century, the Italian physician and researcher Maria Montessori developed the primary educational system that still bears her name. Montessori classrooms emphasize self-directed learning, hands-on engagement with a wide variety of materials (including plants and animals), and a largely unstructured school day. And in recent years they've produced alumni including the founders of Google (Larry Page and Sergey Brin), Amazon (Jeff Bezos), and Wikipedia (Jimmy Wales).

These examples appear to be part of a broader trend. Management researchers Jeffrey Dyer and Hal Gregersen interviewed five hundred prominent innovators and found that a disproportionate number of them also went to Montessori schools, where "they learned to follow their curiosity." As a *Wall Street Journal* blog post by Peter Sims put it, "the Montessori educational approach might be the surest route to joining the creative elite, which are so overrepresented by the school's alumni that one might suspect a Montessori Mafia." Whether or not he's part of this mafia, Andy will vouch for the power of SOLEs. He was a Montessori kid for the earliest years of his schooling, and agrees completely with Larry Page that "part of that training [was] not following rules and orders, and being self-motivated, questioning what's going on in the world, doing things a little bit differently."[12]

Our recommendations about how people can remain valuable knowledge workers in the new machine age are straightforward: work to improve the skills of ideation, large-frame pattern recognition, and complex communication instead of just the three *R*s. And whenever possible, take advantage of self-organizing learning environments, which have a track record of developing these skills in people.

Failing College

Of course, this is easier said than done. And it appears that it's not being done very well in many educational environments. One of the strongest bodies of evidence we've come across that suggests students aren't acquiring the right skills is the work of sociologists Richard Arum and Josipa Roksa and summarized in their book *Academically Adrift: Limited Learning on College Campuses* and subsequent research.[13] Arum and Roksa made use of the Collegiate Learning Assessment (CLA), a recently developed test given to college students to assess their abilities in critical thinking, written communication, problem solving, and analytic reasoning. Although the CLA is administered via computer, it requires essays instead of multiple-choice answers. One of its main components is the 'performance task,' which presents students with a set of background documents and gives them ninety minutes to write an essay requiring them to extract information from the materials given and develop a point of view or recommendation. In short, the performance task is a good test of ideation, pattern recognition, and complex communication.

Arum, Roksa, and their colleagues tracked more than 2,300 students enrolled full-time in four-year degree programs at a range of American colleges and universities. Their findings are alarming: 45 percent of students demonstrate no significant improvement on the CLA after two years of college, and 36 percent did not improve at all even after four years. The average improvement on the test after four years was quite small. Consider a student who scored at the fiftieth percentile as a freshman. If he experienced average improvement over four years of college, then went back and took the test again with another group of incoming freshmen, he would score only in the sixty-eighth percentile. The CLA is so new that we don't know if these gains would have been bigger in the past, but previous research using other tests indicates that they were, and that only a few decades ago the average college student learned a great deal between freshman and senior years.

What accounts for these disappointing results? Arum, Roksa, and their colleagues document that college students today spend only 9 percent of their time studying (compared to 51 percent on "socializing, recreating, and other"), much less than in previous decades, and that only 42 percent reported having taken a class the previous semester that required them to read at least forty pages a week and write at least twenty pages total. They write that, "The portrayal of higher education emerging from [this research] is one of an institution focused more on social than academic experiences. Students spend very little time studying, and professors rarely demand much from them in terms of reading and writing."

They also find, however, that at every college studied some students show great improvement on the CLA. In general, these are students who spent more time studying (especially studying alone), took courses with more required reading and writing, and had more demanding faculty. This pattern fits well into conclusions by education researchers Ernest Pascarella and Patrick Terenzini, who summarized more than twenty years of research in their book *How College Affects Students*. They write that "the impact of college is largely determined by individual effort and involvement in the academic, interpersonal, and extracurricular offerings on a campus."[14]

This work leads directly to our most fundamental recommendation to students and their parents: study hard, using technology and all other available resources to 'fill up your toolkit' and acquire skills and abilities that will be needed in the second machine age.

Tools to Help You Stand Out

Acquiring an excellent education is the best way to not be left behind as technology races ahead. The discouraging news is that today many students seem to be squandering at least some of their educational opportunities. The good news, though, is that technology is now providing more of these opportunities than ever before.

Motivated students and modern technologies are a formidable combination. The best educational resources available online allow users to create self-organized and self-paced learning environments—ones that allow them to spend as much time as they need with the material, and also to take tests that tell them if they mastered it. One of the best known of these resources is Khan Academy, which was started by then–hedge fund manager Salman Khan as a series of online doodles and YouTube video lectures intended to teach math to his young relatives. Their immense popularity led him to quit his job in 2009 and devote himself to creating online educational materials, freely available to all. By May 2013, Khan Academy included more than 4,100 videos, most no more than a few minutes long, on subjects ranging from arithmetic to calculus to physics to art history. These videos had been viewed more than 250 million times, and the Academy's students had tackled more than one billion automatically generated problems.[15]

Khan Academy was originally aimed at primary-school children, but similar tools and techniques have been also applied to higher education, where they're known as massive online open courses, or MOOCs. One of the most interesting experiments in this area came in 2011 when Sebastian Thrun, a top artificial intelligence researcher (and one of the main people behind Google's driverless car), announced with a single email that he would be teaching his graduate-level AI course not only to students at Stanford but also as a MOOC available for free over the Internet. Over 160,000 students signed up for the course. Tens of thousands of them completed all exercises, exams, and other requirements, and some of them did quite well. The top performer in the course at Stanford, in fact, was only the 411th best among all the online students. As Thrun put it, "We just found over 400 people in the world who outperformed the top Stanford student."[16]

In chapter 9, we described the growing gap in earnings between those with and without college degrees. Our MIT colleague David Autor summarizes the research by writing that "large payoffs from schooling are increasingly associated with the attainment of four-year and postcollege degrees. . . . Workers with less than a college education cluster relatively closer together in the earnings distribution while the most educated groups pull away."[17] College graduates are also much less likely to be unemployed than the less educated. Economics reporter Catherine Rampell points out that college graduates are the only group that has seen employment growth since the start of the Great Recession in 2007, and in October of 2011 the unemployment rate for bachelor's degree holders, at 5.8 percent, was only about half that of those with associate's degrees (10.6 percent) and a third that of those who stopped after high school (16.2 percent).[18]

The college premium exists in part because so many types of raw data are getting dramatically cheaper, and as data get cheaper, the bottleneck increasingly is the ability to interpret and use data. This reflects the career advice that Google chief economist Hal Varian frequently gives: seek to be an indispensable complement to something that's getting cheap and plentiful. Examples include data scientists, writers of mobile phone apps, and genetic counselors, who have come into demand as more people have their genes sequenced. Bill Gates has said that he chose to go into software when he saw how cheap and ubiquitous computers, especially microcomputers, were becoming. Jeff Bezos systematically analyzed the bottlenecks and opportunities created by low-cost online commerce, particularly the ability to index large numbers of products, before he set up Amazon. Today, the cognitive skills of college graduates—including not only science, technology, engineering, and math, the so-called STEM disciplines, but also humanities, arts, and social sciences—are often complements to low-cost data and cheap computer power. This helps them command a premium wage.

However, another part of the college premium is less encouraging. More and more employers are requiring college degrees, even for entry-level jobs. As Rampell writes, "The college degree is becoming the new high school diploma: the new minimum requirement, albeit an expensive one, for getting even the lowest-level job. . . . Across industries and geographic areas, many other jobs that didn't used to require a diploma—positions like dental hygienists, cargo agents, clerks and claims adjusters—are increasingly requiring one."[19] This 'degree inflation' is troubling because a college education is expensive and causes many people to go into debt. By the end of 2011, in fact, student loan debt in America was greater than either total outstanding car loans or credit card debt.[20] We hope that MOOCs and other educational innovations eventually provide a lower-cost alternative to traditional colleges, and one that is taken seriously by employers, but until that time comes a college degree remains a vital stepping stone to most careers.

In the future, more and more careers will not be in pure information work—the kind that can be done entirely from a desk. Instead, they will include moving through and interacting with the physical world. This is because computers remain comparatively weak here, even as they get so much stronger at many cognitive tasks.

Advances like autonomous cars, drone airplanes, the Baxter robot, and hacked Kinect devices that can map a room show that great progress has been made in giving machines real-world capabilities, but a towel-folding robot illustrates how far we are from cracking Moravec's paradox. A team of Berkeley researchers equipped a humanoid robot with four stereo cameras and algorithms that would allow it to 'see' towels, both individually and in piles. These algorithms worked; the robot successfully grasped and folded the towels, even though it sometimes took more than one try to grab them correctly. However, it took an average of 1,478 seconds, or more than twenty-four minutes, per towel. The robot spent most of that time looking to learn where the towel was and how to grasp it.[21]

Results like these indicate that cooks, gardeners, repairmen, carpenters, dentists, and home health aides are not about to be replaced by machines in the short term. All of these professions involve a lot of sensorimotor work, and many of them also require the skills of ideation, large-frame pattern recognition, and complex communication. Not all of these jobs are well paying, but they're also not subject to a head-to-head race against the machine.

They may, however, be subject to more competition among people. As the labor market polarizes more and the middle class continues to hollow out, people who were previously doing mid-skill knowledge work start going after jobs lower on the skill and wage ladder. After medical billing specialists have their work automated, for example, they may start looking for jobs as home health aides. This puts downward pressure on wages and makes it harder to find a job in that profession. Even if home health aides remain largely immune to automation, in short, they won't necessarily be immune to all the effects of digitization.

The Fuzzy Future

We have to stress that none of our predictions and recommendations here should be treated as gospel. We don't project that computers and robots are going to acquire the general skills of ideation, large-frame pattern recognition, and highly complex communication any time soon, and we don't think that Moravec's paradox is about to be fully resolved. But one thing we've learned about digital progress is *never say never*. Like many other observers, we've been surprised over and over as digital technologies demonstrated skills and abilities straight out of science fiction.

In fact, the boundary between uniquely human creativity and machine capabilities continues to change. Returning to the game of chess, back in 1956, thirteen-year-old child prodigy Bobby Fischer made a pair of remarkably creative moves against grandmaster Donald Byrne. First he sacrificed his knight, seemingly for no gain, and then exposed his queen to capture. On the surface, these moves seemed insane, but several moves later, Fischer used these moves to win the game. His creativity was hailed at the time as the mark of genius. Yet today if you program that same position into a run-of-the-mill chess program, it will immediately suggest exactly the moves that Fischer played. It's not because the computer has memorized the Fischer–Byrne game, but rather because it searches far enough ahead to see that these moves really do pay off. Sometimes, one man's creativity is another machine's brute-force analysis.[22]

We're very confident that more surprises are in store. After spending time working with leading technologists and watching one bastion of human uniqueness after another fall before the inexorable onslaught of innovation, it's becoming harder and harder to have confidence that any given task will be indefinitely resistant to automation. That means people will need to be more adaptable and flexible in their career aspirations, ready to move on from areas that become subject to automation, and seize new opportunities where machines complement and augment human capabilities. Maybe we'll see a program that can scan the business landscape, spot an opportunity, and write up a business plan so good it'll have venture capitalists ready to invest. Maybe we'll see a computer that can write a thoughtful and insightful report on a complicated topic. Maybe we'll see an automatic medical diagnostician with all the different kinds of knowledge and awareness of a human doctor. And maybe we'll see a computer than can walk up the stairs to an elderly woman's apartment, take her blood pressure, draw blood, and ask if she's been taking her medication, all while putting her at ease instead of terrifying her. We don't think any of these advances is likely to come any time soon, but we've also learned that it's very easy to underestimate the power of digital, exponential, and combinatorial innovation. So never say never.

CHAPTER 13

POLICY RECOMMENDATIONS

"A policy is a temporary creed liable to be changed, but while it holds good it has got to be pursued with apostolic zeal."

—Mahatma Gandhi

WHAT SHOULD WE DO to encourage the bounty of the second machine age while working to reduce the spread, or at least mitigate its harmful effects? How can we best encourage technology to race ahead while ensuring that as few people as possible are left behind?

With so much science-fiction technology becoming reality now every day, it might seem that radical steps are necessary. But this is not the case, at least not right away. Many of the recommendations for growth and prosperity found in just about any standard "Economics 101" textbook are the right place to start and will be for some time to come. In our discussions with policy makers, technologists, and business executives, we were surprised to find that the logic behind these recommendations was often not well understood. Hence this chapter.

A Few Things Even Economists Can Agree On

The standard Econ 101 textbook still provides the right playbook these days because despite recent advances, digital labor is still far from a complete substitute for human labor. Robots and computers, as powerful and capable as they are, are not about to take all of our jobs. Google's autonomous car can't yet drive on all roads or in all conditions, and it doesn't know what to do when a flagman or traffic cop appears in the middle of the street to manually direct traffic.

(That's not to suggest the car would keep driving and run this person over; it would stop and wait for the situation to normalize.) The technologies that make Watson so potent are being applied in many fields, including health care, finance, and customer service, but for now the system is still just a really good *Jeopardy!* player.

In the short term, companies will still need human workers to satisfy their customers and succeed in the economy. (We'll discuss the longer term in the next chapter). Yes, second-machine-age technologies are quickly leaving the lab and entering mainstream business. But as rapid as this progress is, we still have lots of human cashiers, customer service representatives, lawyers, drivers, policemen, home health aides, managers, and other workers. They are not all on the brink of being swept out of their jobs by a cresting wave of computerization. In March 2013 the U.S. workforce consisted of over 142 million people; in each case, their employers chose them over digital technologies (or in addition to them) even after more than fifty years of experience and improvement with business computers, thirty years with PCs, and almost twenty with the World Wide Web.[1] While those employers are likely to choose digital labor more often in the future, it will not be immediate and it will not be in all cases.

For now the best way to tackle our labor force challenges is to grow the economy. As companies see opportunities for growth, the great majority will need to hire people to seize them. Job growth will improve, and so will workers' prospects.

If only growth were that easy. Fierce debates rage about the best ways to bring about faster economic expansion. In particular, there are long-standing and deep disagreements about the proper role of government in this area. Economists, policy makers, and businesspeople alike argue questions of monetary policy—Should the Federal Reserve increase the money supply? What interest should it charge banks?—and fiscal policy—How should the government spend the money it raises? How much debt should it take on? What's the right level and mix of income, sales, corporate, and other taxes? What should the top tax rate be?

Disagreements over these questions often seem so entrenched that there can be no common ground. But there's actually quite a bit of it. Whether you study from the best-selling introductory textbooks *Principles of Economics*, written by Harvard's Greg Mankiw, a conservative economist who advised George Bush and Mitt Romney, or *Economics: An Introductory Analysis*, written by MIT's Paul Samuelson, a liberal advisor to John Kennedy and Lyndon B. Johnson, you'll learn many of the same things.* Across good Econ 101 textbooks, and across good economists, there's far more agreement about government's role in promoting economic growth than you might expect from the more vitriolic public debates in the media. We agree with this Econ 101 playbook as well, and think it will remain central to any appropriate response as machines continue to race ahead.

This playbook advocates government policies and other interventions in a few key areas. Not all of them are concerned with the digital tools of the second machine age. This is because many of the things we should do in a time of brilliant technologies are not related to the technologies themselves. Instead, they're about promoting economic growth and opportunity more generally. Here's our Econ 101 playbook on how to do that.

1. Teach The Children Well

The United States was the clear leader in primary education in the first half of the twentieth century, having realized that inequality was a "race between education and technology," to use a phrase

* The same is true for textbooks by Krugman and Wells, Cowen and Tabarrok, Nordhaus, and on and on.

coined by Jan Tinbergen (winner of the first Nobel Prize in Economic Sciences) and used by the economists Claudia Goldin and Lawrence Katz as the title of their influential 2010 book.[2] When technology advances too quickly for education to keep up, inequality generally rises. Realizing this early last century, the United States made substantial investments in primary education. Goldin documents that by 1955, for example, almost 80 percent of American children between the ages of fifteen and nineteen were enrolled in high schools, a level more than twice as high as that in any European country at the time.

Over the past half century that strong U.S. advantage in primary education has vanished, and the country is now no better than the middle of the pack among wealthy countries, and worse in some important areas. The most recent survey by the Organization for Economic Co-operation and Development's (OECD) Program for International Student Assessment (PISA), conducted in 2009, found that American fifteen-year-olds ranked fourteenth among the thirty-four countries in reading, seventeeth in science, and twenty-fifth in math.[3] As education researcher Martin West summarizes, "In math, the average U.S. student by age 15 was at least a full year behind the average student in six countries, including Canada, Japan, and the Netherlands. Students in six additional countries, including Australia, Belgium, Estonia, and Germany, outperformed U.S. students by more than half a year."[4]

The economic benefits of closing that gap are likely to be quite large. The economists Eric Hanushek and Ludger Woessmann found a strong relationship between improved test scores and faster economic growth after studying forty years' worth of data from fifty countries. This suggests that if the United States could move its students to the top of the international rankings, it might enjoy a substantial boost in GDP growth, especially since many of the country's products and services rely heavily on skilled labor. What's more, it's not an accident that the most educated places in the country, like Austin, Texas; Boston; Minneapolis; and San Francisco have low unemployment rates.

It's been said that America's greatest idea was mass education. It's still a great idea that applies at all levels, not just K-12 and university education, but also preschool, vocational, and lifelong learning.

So, how can we get better results?

USING TECHNOLOGY

We can change the way we deliver education by putting to work digital technologies that have been developed over the past decade or two. The good news is that compared to other industries such as media, retailing, finance, or manufacturing, education is a tremendous laggard in the use of technology. That's good news because it means we can expect big gains simply by catching up to other industries. Innovators can make a huge difference in this area in the coming decade.

The tremendous experimentation now underway with massive online open courses, or MOOCs, is especially encouraging. We discussed MOOCs, which anyone can take, often for free, in some detail in the previous chapter on recommendations for individuals. But we want to point out two of their main economic benefits.

The first and most obvious one is that MOOCs enable low-cost replication of the best teachers, content, and methods. Just as we can all listen to the best pop singer or cellist in the world today, students will soon have access to the most exciting geology demonstrations, the most insightful explanations of Renaissance art, and the most effective exercises for learning statistical techniques. In many cases, we can expect to see schools 'flip the classroom' by having students listen to lectures at home and work through traditional 'homework'—exercises, problem sets, and writing assignments—in school, where peers, teachers, and coaches are available to help them.

The second, subtler benefit from the digitization of education is ultimately more important. Digital education creates an enormous stream of data that makes it possible to give feedback to both teacher and student. Educators can run controlled experiments on teaching methods and adopt a culture of continuous improvement. For instance, one course taught via MITx (MIT's online education initiative) recorded all 230 million times that someone clicked on course materials, and analyzed over 100,000 comments on class discussion boards.[5] The head of MITx, Anant Agarwal, says that he was surprised when the data revealed that half of his students started working on their homework assignments before watching the video lectures. Students were more motivated to really understand the content of the lecture once they saw the specific challenges that they would learn how to overcome.

The real impact of MOOCs is mostly ahead of us, in scaling up the reach of the best teachers, in devising methods to increase the overall level of instruction, and in measuring and finding ways to accelerate student improvement. For millennia teaching methods have remained relatively unchanged: a lone lecturer stands in front of students, working with chalk and slate to illustrate ideas. Our generation is poised to use digitization and analytics to offer a host of improvements. As our friend the technology researcher and professor Venkat Venkatraman put it, "We need digital models of learning and teaching. Not just a technology overlay on old modes of teaching and learning."* We can't predict exactly which methods will be invented and which will catch on, but we do see a clear path for enormous progress. The enthusiasm and optimism in this space is infectious. Given the plethora of new technologies and techniques that are now being explored, it's a certainty that some of them—in

* This was from a posting he put on his Facebook wall—sometimes the medium is part of the message.

fact, we think many of them—will be significant improvements over current approaches to teaching and learning.

A GRAND BARGAIN: HIGHER TEACHER SALARIES AND MORE ACCOUNTABILITY

If there's one consistent finding from educational research, it's that teachers matter. In fact, the impact of a good teacher can be huge. Economists Raj Chetty, John Friedman, and Jonah Rockoff, in a study of 2.5 million American schoolchildren, found that students assigned to better teachers (as measured by their impact on previous students' test scores) earned more as adults, were more likely to attend college, and were less likely to have children as teenagers. They also found that the differences between poor and average teachers can be as important as the ones between average and superior teachers. As they write, "Replacing a [bottom 5 percent] teacher with an average teacher would increase the present value of students' lifetime income by more than $250,000 for the average classroom in our sample."[6]

It seems sensible, then, for educational reforms in the United States to include renewed efforts to attract and retain well-qualified people in the teaching profession, and to remove or retrain consistent low performers.

Part of the bargain should also be longer school hours, longer school years, more after-school activities and more opportunities for preschool education. Studies of successful charter schools by Harvard economist Roland Fryer and others have found that the formula for success is simple, if not easy: longer hours, additional school days, and a no-excuses philosophy that tests students and, implicitly, their teachers.[7] This approach has helped Singapore and South Korea do well in the PISA rankings—both rely heavily on standardized tests for children of all ages.[8] Lengthening the school year may be especially beneficial for poor kids, since research suggests that rich and poor children learn at a similar rate when school is in session, but that poor children fall behind over the summer when they are not in school.[9]

However, one risk of testing is that it can encourage teaching to the test at the expense of other types of learning. We don't necessarily think teaching to the test is always a bad thing, at least for skills that really can be taught and tested, including many basic capabilities that are needed in a global, information-based economy. But it's also important to recognize that hard-to-measure skills like creativity and unstructured problem solving are increasingly important as machines handle more routine work. MIT's Bengt Holmstrom and Stanford's Paul Milgrom did pioneering work showing that strong incentives for achieving measurable goals can crowd out hard-to-measure goals.[10] A clever solution they suggest is via job design and task allocation. Give one group of teachers responsibility for the most measurable goals, while reserving ample time and resources for teachers focusing on the less measurable types of learning, protecting it from being crowded out. In principle, this can achieve the best of both worlds.

We have little doubt that improving education will boost the bounty by providing more of the complementary skills our economy needs to make effective use of new technologies. We're also hopeful that it can help reduce the spread, especially insofar as it's caused by skill-biased technical change. That's largely a matter of supply and demand. Reducing the supply of unskilled workers will relieve some of the downward pressure on their wages, while increasing the supply of educated workers diminishes the shortages in those areas. We also think creativity can be fostered by the right educational settings, boosting the prospects not only of the students but also society as a whole.

But we're also realistic about how new educational technologies are being used in practice. Highly motivated self-starters are the ones who take the greatest advantage of the abundance of online educational resources now available. We know twelve- and fourteen-year-olds who are taking college courses to which they previously would never have had access. Meanwhile, their peers don't participate. Consequently what had been a small gap in their knowledge has become a much larger one. The lesson here is that unless we make real efforts to broaden its impact, the digitization of education won't automatically reduce the spread.

2. Restart Startups

We champion entrepreneurship, but not because we think everyone can or should start a company. Instead, it's because entrepreneurship is the best way to create jobs and opportunity. As old tasks get automated away, along with demand for their corresponding skills, the economy must invent new jobs and industries. Ambitious entrepreneurs are best at this, not well-meaning government leaders or visionary academics. Thomas Edison, Henry Ford, Bill Gates, and many others created new industries that more than replaced the work that was eliminated as farming jobs vanished over the decades. The current transformation of the economy creates an equally large opportunity.

Entrepreneurship has been an important part of the Econ 101 playbook at least since economist Joseph Schumpeter's landmark work, written in the middle of the twentieth century, on the nature of capitalism and innovation. Schumpeter put forward our favorite definition of innovation—"the market introduction of a technical or organisational novelty, not just its invention"—and, like us, believed that it was an essentially recombinant process, "the carrying out of new combinations."[11]

He also argued that innovation was less likely to take place in incumbent companies than in the upstarts that were trying to displace them. As he wrote in *The Theory of Economic Development*, "New combinations are, as a rule, embodied . . . in firms which generally do not arise out of the old ones. . . . It is not the owner of a stage coach who builds railways."[12] Entrepreneurship, then, is an innovation engine. It's also a prime source of job growth. In America, in fact, it appears to be the only thing that's creating jobs. In a study published in 2010, Tim Kane of the Kauffman Foundation used Census Bureau data to divide all U.S. companies into two categories: brand-new startups and existing firms (those that had been around for at least a year). He found that for all but seven years between 1977 and 2005, existing firms as a group were net job destroyers, losing an average of approximately one million jobs annually.[13] Startups, in sharp contrast, created on average a net three million jobs per year.

Subsequent work by John Haltiwanger, Henry Hyatt, and their colleagues confirmed that net job creation is much higher at young companies even though wages are lower.[14] Their research also suggests that startups are responsible for a disproportionate amount of 'worker churn.' This sounds like an unpleasant phenomenon, but it's not; it's mainly workers moving laterally between jobs in search of better opportunities. 'Churn' is an important activity in a healthy economy, but it tends to decrease sharply during recessions, when people become more reluctant to leave their jobs. The group found that young companies increased their share of total churn during the Great Recession and its aftermath, implying that startups provided a much-needed source of transfer opportunities for workers during a difficult period.

America's entrepreneurial environment remains the envy of the rest of the world, but there is troubling evidence that it is becoming less fertile over time. Kauffman Foundation research conducted by economist Robert Fairlie found that while the rate of new business formation rose between 1996 and 2011, most of these startups had a single employee: the founder.[15] This type of entrepreneurship actually increased during the Great Recession, indicating that some entrepreneurs are probably people going into business by themselves after they've lost their jobs. Meanwhile, between 1996 and 2011, the birth rate of 'employer establishments'—companies that employ more than one person at startup—declined by more than 20 percent.

It's not entirely clear what's behind this decline, but the climate facing would-be immigrants might be one factor. In 2012, entrepreneur Vivek Wadhwa and political scientist AnnaLee Saxenian, along with Francis Siciliano, revisited the earlier research they had done on immigrant entrepreneurship. They found that "for the first time in decades, the growth rate of immigrant-founded companies has stagnated, if not declined. In comparison with previous decades of increasing immigrant-led entrepreneurism, the last seven years has witnessed a flattening out of this trend."[16] The change was especially pronounced in Silicon Valley, where over half of companies founded from 1995 to 2005 had at least one immigrant founder. Between 2006 and 2012, that percentage dropped almost ten points, to 43.9 percent.

Another commonly cited culprit behind depressed entrepreneurship is excessive regulation. Innovation researcher Michael Mandel has pointed out that any single regulation might not do much to deter new business formation, but each one is like another pebble in a stream. Their cumulative effect can be increasingly damaging as opportunities to work around them are diminished. There's pretty good evidence that such 'regulatory thickets' are in fact impeding new business formation. For instance, economists Leora Klapper, Luc Laeven, and Raghuram Rajan found that higher levels of regulation reduce startup activity.[17] Their research was conducted using European data, but it seems likely that its conclusions are at least in part applicable to the United States as well.

We favor reducing unnecessary, redundant, and overly burdensome regulation, but recognize that this is likely to be slow and difficult work. First, regulators rarely like giving up authority once it's granted to them. Second, those companies and industries protected by existing regulations will no doubt lobby strenuously to preserve their privileged positions. And third, separate sets of regulations exist at the federal, state, and municipal levels in America, so comprehensive change cannot be brought about by any single entity. The country's Constitution is clear that most powers related to commerce rest with the individual states, so prospective entrepreneurs can likely expect to face a continued patchwork of regulations in many areas. Still, we believe that it is important to try to reduce the regulatory burden and make the business environment as welcoming as possible for entrepreneurs.

We don't expect anyone to duplicate Silicon Valley, but we do think government, businesses, and individuals can do more to fuel high-growth entrepreneurship. An intriguing example is the work that Steve Case and the Kauffman Foundation are doing with the Startup America Partnership. It seeks to support over thirty entrepreneur-led startup regions, complete with a 'dating site' to make it easier for new ventures to partner with Fortune 500 firms that can complement their innovations with marketing, manufacturing, or distribution networks.

3. Make More Matches

Although job sites like Monster.com and Aftercollege.com and networking sites like LinkedIn have made it easier for employers and employees to find one another, the vast majority of our students that graduate each year still rely primarily on word of mouth recommendations from friends, relatives, and, yes, professors, to make introductions. We must find ways to reduce the friction and search costs that make it unnecessarily difficult to match people with jobs.

LinkedIn is developing a real-time database that describes the skills sought by companies and matches those skills with the training that students and other potential employees have. Sometimes simply rewording a few concepts on a resume can make the difference: companies looking for app developers for Android phones, for example, may not realize that a software development class on a student's resume used that operating system.

Local, national, and global databases of job opportunities and candidates can have a huge payoff. Too often employers focus narrowly on graduates from a few schools when there are thousands of equal or better-qualified candidates. The federal government could use prizes to spur development of these databases. We should also encourage and support private companies to develop better algorithms and techniques for identifying skills and matching them to employers. For instance, a company called Knack, which Erik advises, has developed a series of games, each of which generates megabytes of data. By mining the data, Knack can get surprisingly accurate assessments of the players' creativity, persistence, extroversion, diligence, and other characteristics that are hard to discern from a college transcript or even a face-to-face interview. Other companies such as HireArt and oDesk are also using analytics to create better matches and less friction in the employment market. We are also encouraged by the burgeoning use of ratings like TopCoder scores to provide objective metrics of candidate skills. This makes it easier for job seekers to find their best niches and for entrepreneurs and employers to find the talent they need.

4. Support Our Scientists

After rising for a quarter-century, U.S. federal government support for basic academic research started to fall in 2005.[18] This is cause for concern because economics teaches that basic research has large beneficial externalities. This fact creates a role for government, and the payoff can be enormous. The Internet, to take one famous example, was born out of U.S. Defense Department research into how to build bomb-proof networks. GPS systems, touchscreen displays, voice recognition software like Apple's Siri, and many other digital innovations also arose from basic research sponsored by the government. It's pretty safe to say, in fact, that hardware, software, networks, and robots would not exist in anything like the volume, variety, and forms we know today without sustained government funding.[19] This funding should be continued, and the recent dispiriting trend of reduced federal funding for basic research in America should be reversed.

We should also reform the U.S. intellectual property regime, particularly when it comes to software patents and copyright duration. In any age, but especially in the second machine age, intellectual property is extremely important. It's both a reward for innovation (if someone invents a better mousetrap, he or she gets to patent it) and an input to it (most new ideas are recombinations of existing ones). Governments therefore have to strike a delicate balance; they have to provide enough intellectual property protection to encourage innovation but not so much that they stifle it. Many of today's informed observers conclude that software patents are providing too much protection. The same is probably true for at least some copyrights; it's not clear what public interest is served by laws that ensure Disney's 1928 "Steamboat Willie" (precursor to Mickey Mouse) remains under copyright, as does the song "Happy Birthday."[20]

PRIZES

Many innovations are of course impossible to describe in advance (that's what makes them innovations). But there are also cases where we know exactly what we're looking for and just want somebody to invent it. In these cases, prizes can be especially effective.* Google's driverless car was a direct outgrowth of a Defense Advanced Research Projects Agency (DARPA) challenge that offered a one-million-dollar prize for a car that could navigate a spe-

* Prizes have a long history going back to the Longitude Prize offered by act of the British Parliament in 1714. While latitude was relatively easy to calculate, longitude was a bigger problem, especially during long ocean voyages. A series of prizes totaling over one hundred thousand British pounds motivated major advances throughout the 1700s in the measurement of longitude. In 1919, the twenty-five-thousand-dollar Orteig Prize for a nonstop transatlantic flight motivated a series of aviation innovations, culminating in Charles Lindbergh's successful flight in 1927.

cific course without a human driver. Tom Kalil, Deputy Director for Policy of the United States Office of Science and Technology Policy, provides a great playbook for how to run a prize:[21]

1. Shine a spotlight on a problem or opportunity
2. Pay only for results
3. Target an ambitious goal without predicting which team or approach is most likely to succeed
4. Reach beyond usual suspects to tap top talent
5. Stimulate private-sector investment many times greater than the prize purse
6. Bring out-of-discipline perspectives to bear
7. Inspire risk taking by offering a level playing field
8. Establish clear target metrics and validation protocols

Over the past decade, the total federal and private funds earmarked for large prizes have more than tripled and now surpass $375 million.[22] This is great, but it's just a tiny fraction of overall government spending on research. There remains great scope for increasing the volume and variety of innovation competitions.

5. Upgrade Infrastructure

It's almost universally agreed among economists that the government should be involved in building and maintaining infrastructure—streets and highways, bridges, ports, dams, airports and air traffic control systems, and so on. This is because, like education and research, infrastructure is subject to positive externalities.

Excellent infrastructure makes a country a more pleasant place to live, and also a more productive place in which to do business. Ours, however, is not in good shape. The American Society of Civil Engineers (ASCE) gave the United States an overall infrastructure grade of D+ in 2013, and estimated that the country has a backlog of over $3.6 trillion in infrastructure investment.[23] However, only a bit more than $2 trillion has been budgeted to be spent by 2020, leaving a large gap. You might think that the ASCE has an obvious bias on the question of infrastructure spending, but the data bear them out. Between 2009 and 2013, public investment in infrastructure fell by over $120 billion in real terms, to its lowest level since 2001.[24]

Bringing U.S. infrastructure up to an acceptable grade would be one of the best investments the country could make in its own future. As we write in 2013, energy prices are dropping, thanks in large part to the domestic shale oil boom, and wages in countries like China are rising. Because of these and other factors, we often hear from business leaders something very close to what Eric Spiegel, the CEO of Siemens USA, said in an interview: "The U.S. is a great place for manufacturing these days. We're making things here in the U.S. that we're shipping over to China. . . . We just need to make sure that we've . . . got the infrastructure in place to be able to handle the increased work."[25]

There's an interesting historical wrinkle in discussions about infrastructure investment. The legendary economist John Maynard Keynes, whose name is attached to a school of thought that advocates stimulus spending, famously suggested in 1936 that during recessions the government should put money in bottles, bury the bottles deep in old coal mines, then sell the rights to dig them up.[26] Doing so, he argued only partly in jest, would "be better than nothing" because it would create demand during periods when labor and capital would otherwise go unused. Economists fiercely debate whether or not this could actually work, but they rarely debate the merits of good roads and bridges, or of government involvement with them because of positive externalities. We're making our argument for infrastructure investment because of these externalities, independent of any Keynesian stimulus it might provide, and we're squarely in the economic mainstream when we do so.

WELCOME THE WORLD'S TALENT

Any policy shift advocated by both the libertarian Cato Institute and the progressive Center for American Progress can truly be said to have diverse support.[27] Such is the case for immigration reform, a range of proposed changes with the broad goal of increasing the numbers of legal foreign-born workers and citizens in the United States. Generous immigration policies really are part of the Econ 101 playbook; there is wide agreement among economists that they benefit not only the immigrants themselves but also the economy of the country they move to.

Some studies have found that certain workers in the host country, particularly less skilled ones, are made worse off by immigration because their wages fall but other research has reached different conclusions. Economist David Card, for example, evaluated the impact of Cuba's 1980 Mariel boatlift (a mass emigration of Cubans to the United States approved by Fidel Castro) on the Miami labor market. Mariel brought over one hundred thousand people to the city in less than a year and increased its labor force by 7 percent, yet Card found "virtually no effect on the wages or unemployment rates of less-skilled workers, even among Cubans who had immigrated earlier."[28] Economist Rachel Friedberg reached virtually the same conclusion about mass migration from Russia and the rest of the former Soviet Union into Israel.[29] Despite increasing the country's population by 12 percent between 1990 and 1994, this immigration had no discernible adverse effect on Israeli workers.

Despite this and other evidence, concerns persist in America that large-scale immigration of unskilled workers, particularly from Mexico and other Latin American countries and particularly by illegal means, will harm the economic prospects of the native-born labor force. Since 2007, it appears that net illegal immigration to the United States is approximately zero, or actually negative.[30] And a study by the Brookings Institution found that highly educated immigrants now outnumber less educated ones; in 2010, 30 percent had at least a college education, while only 28 percent lacked the equivalent of a high school degree.[31]

Entrepreneurship in America, particularly in technology-intensive sectors of the economy, is fueled by immigration to an extraordinary degree. Foreign-born people make up less than 13 percent of the country's population in recent years, but between 1995 and 2005 more than 25 percent of all new engineering and technology companies had at least one immigrant cofounder, according to research by Wadhwa, Saxenian, and their colleagues.[32] These companies in total had more than $52 billion in 2005 sales, and employed almost 450,000 people. According to immigration reform advocacy group Partnership for a New American Economy, between 1990 and 2005, 25 percent of America's highest-growth companies were founded by foreign-born entrepreneurs.[33] As economist Michael Kremer demonstrated in a now classic paper, increasing the number of immigrant engineers actually leads to higher, not lower, wages for native-born engineers because immigrants help creative ecosystems flourish.[34] It's no wonder that wages are higher for good software designers in Silicon Valley, where they are surrounded by others with similar and generally complementary skills, rather than in more isolated parts of the world.

Today, immigrants are having this large and beneficial effect on the country not because of America's processes and policies but often despite them. Immigration to the United States is often described as slow, complex, inefficient, and highly bureaucratic. Darrell West, a vice president at the Brookings Institution, wrote a book in 2011 called *Brain Gain: Rethinking U.S. Immigration Policy*. But his research didn't prepare him for his own Kafkaesque experiences after he married a German woman who then sought American citizenship. He wrote, "For many immigrants, it is virtually impossible for them to afford the fees, handle the paperwork, and navigate a complex bureaucratic process. Even with a Ph.D. in political science,

I was overwhelmed with the complexity of the multiple applications, fees, documentation, interviews, and trips to the immigration office. . . . American immigration is a 19th century process in a 21st century world."[35]

In addition to broken processes, the United States also has counterproductive immigration policies. Among technologists, the clearest example of this is probably the annual cap on the number of H1-B visas issued. These allow U.S. employers to hire foreign workers in specialty occupations, usually technical, for up to six years. In the early years of the twenty-first century as many as 195,000 were issued annually, but the quota was reduced to 65,000 in 2004 (in 2006 the program was expanded to include 20,000 graduates of American universities).

The H1-B visa program should be further expanded. We like the imagery of stapling a green card to every advanced diploma awarded to an immigrant. We also support the creation of a separate 'startup visa' category aimed at making it easier for entrepreneurs, especially those who have already attracted funding, to launch their ventures in the United States. This idea has been championed most prominently by American venture capitalists and business groups, but other countries have taken the lead. Australia, the UK, and Chile have all launched programs to attract early-stage entrepreneur immigrants, and in January 2013 Canada announced a full-fledged startup visa program, the first of its kind in the world.[36] Meanwhile, comprehensive immigration reform stalled in the U.S. Congress in the summer of that same year.

6. Since We Must Tax, Tax Wisely

In general, taxing something discourages its production. That's usually considered a bad thing, but it doesn't have to be since we can tax things we want less of. There are also some goods and services that are exceptions to the rule; taxation doesn't lead to decreases in the amount of them available. Economists say that these offerings are provided inelastically with respect to taxation. We can and should take advantage of this fact.

PIGOVIAN TAXES

A factory might find it really cheap and convenient to dump all of its waste into the river that flows past it, but the resulting toxic water, dead fish, and nasty smell are clearly unwanted. Economists call this type of unwanted effect a *negative externality*. Many types of pollution are prohibited outright as a result, but it's not possible or smart to forbid every type. Utilities have to generate some pollution when they generate electricity, for example, and while cars today run much more cleanly than they used to, they still give off greenhouse gases. It is an unfortunate fact of human life that some types of production generate 'bads' alongside goods.

In cases like these, most economists advocate taxing the pollution. Such taxes are called "Pigovian" after Arthur Pigou, a British economist of the early twentieth century who was one of their early champions. The taxes have two important benefits. First, they reduce the amount of undesirable activity; if a utility gets taxed based on the amount of sulfur dioxide it releases into the atmosphere, it has strong incentives to invest in scrubber technology that leaves the air cleaner. Second, Pigovian taxes raise revenue for the government, which could be used to compensate those harmed by the pollution (or for any other purpose). They're a win-win. Taxes of this type are popular across the political spectrum and among people in many fields; members of the "Pigou Club," a group of advocates identified by economist Gregory Mankiw, include both Alan Greenspan and Ralph Nader.[37]

By improving measurement and metering, the technologies of the second machine age make Pigovian taxes more feasible. Consider traffic congestion. Each of us imposes a cost on all other drivers when we join an already overcrowded highway and further slow traffic. At peak hours, traffic on Interstate 405 in Los Angeles crawls at fourteen miles per hour, more than quadrupling what should be an eight-minute drive. Congestion pricing, aided by electronic passes or digital cameras, can dynamically adjust the cost of the roadway so that drivers would only choose to drive when the total cost created, including the additional congestion, was less than the value of their trip.

Congestion-reducing activities like carpooling, off-peak commuting, bicycling, telecommuting, and mass transit would all increase with congestion pricing in effect. Already Pigovian principles have been applied to revenue-generating segments of infrastructure like toll roads and London's congestion zone, which reduces traffic and takes in money by charging motorists to drive into the city center during peak times. Meanwhile, Singapore has implemented an Electronic Road Pricing System that has virtually eliminated congestion.

Americans collectively spend over one hundred billion hours stuck in traffic jams, a testament to the fact that road pricing is not yet widely adopted. By some estimates, the revenues from optimal congestion pricing would be enough to eliminate all state taxes in California. In the past, it was impossible to meter road usage in a cost-effective way, so we settled for leaving it unpriced and putting up with what resulted: the kinds of long lines and waiting we rarely saw outside the former Soviet Union for other goods and services. Digital road pricing systems could help us recapture that lost time while replacing revenues from other sources.

TAXES ON ECONOMIC RENTS

The supply of some goods, like land, is completely inelastic—there's the same amount of land, no matter how heavily it's taxed. That means that a tax on the revenues from that good (in other words the 'economic rents' from it) will not reduce its supply. As a result, such taxes are relatively efficient—they don't distort incentives or activities. The nineteenth-century economist Henry George took this insight and argued that we should have just a single tax, a land tax. While an enticing idea, the reality is that revenues from land rents aren't high enough to pay for all government services. Still, they could pay for more than they currently do, and there are other rents in the economy, including those from natural resources like government-owned oil and gas leases, that could be significantly increased.

There's also an argument that a big part of the very high earnings of many 'superstars' are also rents. These questions turn on whether most professional athletes, CEOs, media personalities, or rock stars are genuinely motivated by the absolute level of their compensation versus the relative compensation, their fame, or their intrinsic love of their work. In all likelihood, we could raise more revenue by increasing marginal tax rates on the highest income earners, for instance by introducing new tax brackets at the one-million- and ten-million-dollar levels of annual income. We do not find much evidence supporting the counterargument that higher taxes on this population will harm economic growth by eroding high earners' initiative. In fact, research by our MIT colleague and Nobel Prize–winning economist Peter Diamond, in partnership with Clark Medal winner Emmanuel Saez, suggests that optimal tax rates at the very top of the income distribution might be as high as 76 percent.[38] While we don't see the need for that level of taxation, we do take comfort from the fact that the last time income taxes were substantially raised under Bill Clinton, the economy grew rapidly in the years that followed. Indeed, as noted by economist Menzie Chinn, there is no visible relationship between top tax rates and overall economic growth, at least in the ranges the U.S. experienced.[39]

We don't pretend that the policies we advocate here will be easy to adopt in the current political climate, or that if they somehow were all adopted they would immediately bring back full employment and rising average wages. We know that these are challenging times; many people have seen their fortunes suffer during the Great Recession and subsequent slow recovery and are being left behind by the twin forces of technology and globalization. Inequality and other forms of spread are increasing, and everyone is not sharing in all the types of bounty the economy is generating.

The policy recommendations we outline above share one simple and modest goal: bringing about higher rates of overall economic growth. If this happens, the prospects of workers and job seekers alike will improve.

CHAPTER 14

LONG-TERM RECOMMENDATIONS

"Work saves a man from three great evils: boredom, vice, and need."

—Voltaire

THE RECOMMENDATIONS WE MADE in the previous chapter will help boost the bounty and reduce or reverse the spread. But as we move deeper into the second machine age and the second half of the chessboard, will the Econ 101 playbook be enough to maintain healthy wage and job prospects?

As we look further ahead—into the 2020s and beyond—we see androids. They don't look like the machines in the *Matrix* or *Terminator* movies—some don't even have physical bodies; they're not going to declare war on us, and they're not going to replace all human workers, or even most of them, in the next few years. But as we've seen in earlier chapters, technology is steadily encroaching on humans' skills and abilities. So what should we do about the fact that the androids are coming? What are the right policies and interventions going forward?

Please, No Politburos

Let's start by being humble. History is littered with unintended and sometimes tragic side effects of well-intentioned social and economic policies. It's difficult to know in advance exactly which changes will be most disruptive, which will be implemented with unexpected ease, and how people will react in an environment that has never before been observed.

Caveats aside, we do have some ideas about how to proceed, and how not to. We do not think the right policy would be to try to halt the march of technology, or to somehow disable the mix of exponential, digital, combinatorial innovation taking place at present. Doing so would be about as bad an idea as locking all the schools and burning all the scientific journals. At best, such moves would ensure the status quo at the expense of betterment or progress. As the technologist Tim O'Reilly puts it, they'd be efforts to protect the past against the future.[1] So would attempts to protect today's jobs by short-circuiting tomorrow's technologies. We need to let the technologies of the second machine age do their work and find ways to deal with the challenges they will bring with them.

We are also skeptical of efforts to come up with fundamental alternatives to capitalism. By 'capitalism' here, we mean a decentralized economic system of production and exchange in which most of the means of production are in private hands (as opposed to belonging to the government), where most exchange is voluntary (no one can force you to sign a contract against your will), and where most goods have prices that vary based on relative supply and demand instead of being fixed by a central authority. All of these features exist in most economies around the world today. Many are even in place in today's China, which is still officially communist.

These features are so widespread because they work so well. Capitalism allocates resources, generates innovation, rewards effort, and builds affluence with high efficiency, and these are extraordinarily important things to do well in a society. As a system capitalism is not perfect, but it's far better than the alternatives. Winston Churchill said that, "Democracy is the worst form of government except for all those others that have been tried."[2] We believe the same about capitalism.

The element that's most likely to change, and to present challenges, is one that we have not mentioned yet: in today's capitalist economies, most people acquire money to buy things by offering their labor to the economy. Most of us are laborers, not owners of capital. If our android thought experiment is correct, though, this long-standing exchange will become less feasible over time. As digital labor becomes more pervasive, capable, and powerful, companies will be increasingly unwilling to pay people wages that they'll accept and that will allow them to maintain the standard of living to which they've been accustomed. When this happens, they remain unemployed. This is bad news for the economy, since unemployed people don't create much demand for goods and overall growth slows down. Weak demand can lead to further deterioration in wages and unemployment as well as less investment in human capital and in equipment, and a vicious cycle can take hold.

Revisiting the Basic Income

A number of economists have been concerned about this possible failure mode of capitalism. Many of them have proposed the same simple solution: give people money. The easiest way to do this would have the government distribute an equal amount of money to everyone in the country each year, without doing any means of testing or other evaluation of who needs the money or who should get more or less. This 'basic income' scheme, its proponents argue, is comparatively straightforward to administer, and it preserves the elements of capitalism that work well while addressing the problem that some people can't make a living by offering their labor. The basic income assures that everyone has a minimum standard of living. If people want to improve on it by working, investing, starting a company, or doing any of the other activities of the capitalist engine they certainly can, but even if they don't they will still be able to act as consumers, since they will still receive money.

Basic income is not part of mainstream policy discussions today, but it has a surprisingly long history and came remarkably close to reality in twentieth-century America. One of its early proponents was the English-American political activist Thomas Paine, who advocated in his 1797 pamphlet *Agrarian Justice* that everyone should be given a lump sum of money upon reaching adulthood to compensate for the unjust fact that some people were born into landowning families while others were not. Later advocates included philosopher Bertrand Russell and civil rights leader Martin Luther King, Jr., who wrote in 1967, "I am now convinced that the simplest approach will prove to be the most effective—the solution to poverty is to abolish it directly by a now widely discussed measure: the guaranteed income."[3]

Many economists on both the left and the right have agreed with King. Liberals including James Tobin, Paul Samuelson, and John Kenneth Galbraith and conservatives like Milton Friedman and Friedrich Hayek have all advocated income guarantees in one form or another, and in 1968 more than 1,200 economists signed a letter in support of the concept addressed to the U.S. Congress.[4]

The president elected that year, Republican Richard Nixon, tried throughout his first term in office to enact it into law. In a 1969 speech he proposed a Family Assistance Plan that had many features of a basic income program. The plan had support across the ideological spectrum, but it also faced a large and diverse group of opponents.[5] Caseworkers and other administrators of existing welfare programs feared that their jobs would be eliminated under the new regime; some labor leaders thought that it would erode support for minimum wage legislation; and many working Americans didn't like the idea of their tax dollars going to people who could work, but chose not to. By the time of his 1972 reelection campaign, Nixon had abandoned the Family Assistance Plan, and universal income guarantee programs have not been seriously discussed by federal elected officials and policymakers since then.*

* The state of Alaska, however, set up a form of guaranteed income for its residents in 1980, when it passed legislation establishing universal dividends from its Permanent Fund. The Fund was set up in 1976 to manage the state's share of its abundant oil wealth; four years later, Alaskans decided

Avoiding the Three Great Evils

Will we need to revisit the idea of a basic income in the decades to come? Maybe, but it's not our first choice. Voltaire beautifully summarized why not when he made the observation quoted at the start of this chapter: "Work saves a man from three great evils: boredom, vice, and need."[6] A guaranteed universal income takes care of need, but not the other two. And just about all the research and evidence we've looked at has convinced us that Voltaire was right. It's tremendously important for people to work not just because that's how they get their money, but also because it's one of the principal ways they get many other important things: self-worth, community, engagement, healthy values, structure, and dignity, to name just a few.

Whether the focus is on the individual or the community, the conclusion is the same: work is beneficial. At the individual level there has been a great deal of research into what makes people feel fulfilled, content, and happy. In his book *Drive*, Daniel Pink summarizes the three key motivations from the research literature: mastery, autonomy, and purpose.[7] The last of these was emphasized by an older worker quoted in a February 2013 story about the pros and cons of the warehouse jobs online retail giant Amazon was creating in the UK: "It gives you your pride back. That's what it gives you. Your pride back."[8] His view is strongly supported by the work of economist Andrew Oswald, who found that joblessness lasting six months or longer harms feelings of well-being and other measures of mental health about as much as the death of a spouse, and that little of this decline is due to the loss of income; instead, it arises from a loss of self-worth.[9]

A survey of people in many countries conducted by the Gallup

---

that a portion of this wealth should be distributed each year in the form of dividend checks.

polling organization confirmed the fundamental desire for work. As Gallup CEO Jim Clifton puts it in his book *The Coming Jobs War*, "The primary will of the world is no longer about peace or freedom or even democracy; it is not about having a family, and it is neither about God nor about owning a home or land. The will of the world is first and foremost to have a good job. Everything else comes after that."[10] It seems that all around the world, people want to escape the evils of boredom, vice, and need and instead find mastery, autonomy, and purpose by working.

A lack of work harms not just individuals but entire communities. Sociologist William Julius Wilson summarized a long career's worth of findings in his 1996 book *When Work Disappears*. His conclusions are unequivocal:

> The consequences of high neighborhood joblessness are more devastating than those of high neighborhood poverty. A neighborhood in which people are poor but employed is different from a neighborhood in which many people are poor and jobless. Many of today's problems in the inner-city ghetto neighborhoods—crime, family dissolution, welfare, low levels of social organization, and so on—are fundamentally a consequence of the disappearance of work.[11]

In his 2012 book *Coming Apart*, social researcher Charles Murray put numbers to the problems Wilson described and also showed that they weren't confined to inner cities or largely minority neighborhoods. Instead, they were squarely part of mainstream white America. Murray identified two groups. The first comprises Americans with at least a college education and a professional or managerial job; these are dubbed residents of the hypothetical town 'Belmont,' named after a prosperous suburb of Boston. The second group consists of those with no more than a high school education and a blue-collar or clerical job; these are residents of 'Fishtown,' named after a working-class suburb of Philadelphia. In 2010 approximately

30 percent of the American workforce lived in Belmont, 20 percent in Fishtown.[12]

Using a variety of data sources, Murray tracked what happened in Belmont and Fishtown from 1960 to 2010. At the start of that time span the two towns were not that far apart in most measures that track the health of a community—marriage, divorce, crime, etc.—and they were also both full of people that worked. In 1960, 90 percent of Belmont households had at least one adult working forty or more hours a week, as did 81 percent of Fishtown households. By 2010 the situation had changed drastically for one of the communities. While 87 percent of Belmont households still had at least one person working that much, only 53 percent of Fishtown households did.

What else changed in Fishtown? Many things, none of them good. Marriages became less happy, and less common. In 1960, only about 5 percent of Fishtowners between the ages of thirty and forty-nine were divorced or separated; by 2010, a third of them were. Over time, many fewer children in Fishtown grew up in two-parent homes; by 2004, the figure had dropped below 30 percent. And incarceration rates skyrocketed; in 1974, 213 out of every 100,000 Fishtowners were in prison. That number grew more than fourfold, to 957, over the next thirty years. Belmont also saw negative changes in some of these areas, but they were tiny in comparison. As late as 2004, for example, fully 90 percent of children in Belmont were still living with both of their biological parents.

The disappearance of work was not the only force driving Belmont and Fishtown apart—Murray himself focuses on other factors[13]—but we believe it is a very important one. The evidence suggests that communities in which people are working are much healthier than communities where work is scarce, all other things being equal. So we support policies that encourage work, even as the second machine age progresses.

And we see two pieces of good news here. The first is that economists have developed interventions that encourage and reward work in ways that a basic income guarantee alone does not. The second is that innovators and entrepreneurs have developed technologies not only to substitute for human labor but also to complement it. In other words, digital tools are not just taking work out of the economy; they're also providing new opportunities for people to contribute work to it. As technology keeps racing ahead the best approach is to combine these two pieces of good news and try to maintain an economy of workers. Doing so will address all three of Voltaire's evils and give us a much better chance of maintaining not only a bounteous economy, but also a healthy society.

Better Than Basic: The Negative Income Tax

The Nobel Prize–winning conservative economist Milton Friedman did not advocate many government interventions, but he was in favor of what he termed a 'negative income tax' to help the poor. As he explained it in a 1968 television appearance:

> Under present law we have a positive Income Tax that everybody knows about. . . . [U]nder the Positive Income Tax if you happen to be the head of a family of four, for example, and you have $3,000 of income, you neither pay a tax nor receive any benefit from it. You're just on the break-even point. Suppose you have an income of $4,000. Then you have $1,000 of positive taxable income, on which at current rates (14%) you pay $140.00 in tax. Suppose today you had an income of $2,000. Well then you're entitled to deductions and exemptions of $3,000, you have an income of $2,000. You have a negative . . . taxable income of $1,000. But currently under present law you get no benefit of those unused deductions. The idea of a Negative Income Tax is that, when your income is below the break-even point, you would get a fraction of it as a payment "from" the government. You would receive the funds instead of paying them.[14]

To finish his example, if the negative income tax rate were 50 percent, the person making $2,000 would get $500 back from the government, which is $1,000 (the negative taxable income) times .50 (the 50-percent negative income tax rate), and would thus have total income for that year of $2,500. A person with zero income would get $1,500 from the government, since they had $3,000 of negative taxable income.

The negative income tax combines a guaranteed minimum income with an incentive to work. Below the cutoff point in the example (which was $3,000 in 1968 but would be about $20,000 in 2013 dollars), every dollar earned still increases total income by $1.50. This encourages people to start working and keep finding more work to do, even if the wages they receive for this work are low. It also encourages them to file tax returns and so become part of the visible mainstream workforce. In addition, it is relatively straightforward to administer, making use of the existing infrastructure for filing taxes and distributing refunds.

For all these reasons, we like the idea of a negative income tax. At present, the American federal tax system includes a related idea called the Earned Income Tax Credit, or EITC. Compared to Freidman's forty-year-old proposal, however, the EITC is small; in 2012 it maxed out at less than $6,000 for families with three or more qualifying children and less than $500 for families with no children. In addition, it cannot be used by people who have no income. Even though it's small, though, the EITC is still powerful: research by economists Raj Chetty and Nathaniel Hendren at Harvard, along with Patrick Kline and Emmanuel Saez at Berkeley, suggests that states with more generous EITC policies also have significantly greater intergenerational mobility.[15]

We support turning the EITC into a full-fledged negative income tax by making it larger and making it universal. We also think claiming the EITC should be made easier and more obvious. Approximately 20 percent of eligible taxpayers don't take advantage of it, probably because they aren't aware of its existence or are put off by its complexity.[16]

The EITC is really a subsidy on labor, paying a bonus dollar of labor income. It puts into practice some of the oldest economic advice of all: tax things you want to see less of, and subsidize things you want to see more of. We tax cigarettes and gas-guzzling cars, for example, and subsidize solar panel installations.[17] The idea, of course, is that the tax will decrease the incidence of the undesirable activity (smoking cigarettes, driving a gas-guzzler) by making it more expensive, while the subsidy will have exactly the opposite effect. We agree with our MIT colleague Tom Kochan, who thinks of unemployment as a kind of "market failure," or externality. That means that the benefits of increasing employment—reduced crime, greater investment, and stronger communities—extend to people throughout society, not just the employer or employee who are party to the employment contract. If unemployment creates negative externalities, then we should reward employment instead of taxing it.

It's not always possible to follow this advice. The U.S. government taxes labor not because it wants people to be idle but because it needs to raise money somehow, and income and labor taxes have historically been the preferred method. The income tax first appeared during the Civil War and was made permanent in 1913 by the Sixteenth Amendment to the Constitution.[18] By 2010, over 80 percent of all revenue collected by the federal government came from individual income taxes and payroll taxes. In turn, payroll taxes fall into two categories. The first are payroll taxes withheld by employers from their employees' wages; the second are per-employee taxes charged to the employers themselves. Payroll taxes, which fund programs like Medicare, Social Security, and unemployment insurance, accounted for only about 10 percent of federal tax revenue early in the 1950s but now make up about 40 percent, an amount roughly equal to that raised by the individual income tax.[19]

While income taxes are not meant to discourage work and employment, they can still have this effect. Payroll taxes can lead to similar shifts, and by design mainly affect people with low and middle incomes.[20] They can cause organizations to move away from hiring additional domestic employees, and instead outsource work or make use of part-time contractors. As digital technologies keep acquiring new skills and capabilities, these same organizations will increasingly have another option: they'll be able to make use of digital laborers rather than humans. The more expensive human labor is, the more readily employers will switch over to machines. And since payroll taxes make human labor more expensive, they'll very likely have the effect of hastening this switch. Mandates like employer-provided health care coverage have the same effect; they too appear as a tax on human labor and so discourages it, all other things being equal.[21]

We bring up these points not because we dislike Social Security or health care coverage. We like both of them a great deal and want them to continue. We simply point out that these and other popular programs are financed, in whole or in part, by taxes on labor. This might have been an appropriate idea when there were no viable alternatives to humans for most jobs, but that is no longer the case. The better machines become at substituting for human labor, the bigger negative effect any tax or mandate will have on human employment.

So in addition to subsidizing work via a negative income tax, we also support not taxing work as much in the first place and reducing burdens and mandates on employers. Like so much else at the intersection of economics and policy, this is easy to say and extremely hard to enact. How else, if not by taxes on labor, are expensive, popular, and important programs like Social Security and Medicare to be funded? How is health care coverage to be provided if not by employers?

We don't claim to have all the answers to these critical questions, but we do know that the economist's toolkit contains other kinds of taxes besides those on labor. As discussed in the last chapter, these include Pigovian taxes on pollution and other negative externalities, consumption taxes, and the value-added tax (VAT), which companies pay based on the difference between their costs (labor, raw materials, and so on) and the prices they charge customers. A VAT has several attractive properties—it's relatively straightforward to collect, adjustable, and lucrative—but is not currently used in the United States. In fact, America is the only one of the thirty-four nations in the OECD without one. Economist Bruce Bartlett, legal scholar Michael Graetz, and others have put together alternatives to the current American tax system that rely heavily on a VAT.[22] We think these are valuable contributions to the discussion about how to best pay for government services in the second machine age, and deserve serious consideration.

The Peer Economy and Artificial Artificial Intelligence

Changing the subsidies and taxes on labor might seem like a short-term solution. After all, isn't the second machine age defined by relentless automation that will lead to a largely or completely post-work economy?

We've argued here that in many domains it is. But, as we've also hopefully shown, people have skills and abilities that are not yet automated. They may become automatable at some point but this hasn't started in any serious way thus far, which leads us to believe that it will take a while. We think we'll have human data scientists, conference organizers, divisional managers, nurses, and busboys for some time to come.

And as we discussed previously, people still have much to offer even in heavily automated domains. Although no person now can beat the best chess computer, for example, the right mix of human and digital labor easily beats it. So it's not the case that people cease to be valuable the instant computers surpass them in a domain.

They can be enormously useful once they've paired up to race *with* machines, instead of against them.

We see this even in heavily automated fields like computer search. As Steve Lohr explained in a March 2013 *New York Times* story,

> [W]hen Mitt Romney talked of cutting government money for public broadcasting in a presidential debate last fall and mentioned Big Bird, [Twitter] messages with that phrase surged. Human judges recognized instantly that "Big Bird," in that context and at that moment, was mainly a political comment, not a reference to "Sesame Street," and that politics-related messages should pop up when someone searched for "Big Bird." People can understand such references more accurately and quickly than software can, and their judgments are fed immediately into Twitter's search algorithm. . . .
>
> Other human helpers, known as evaluators or raters, help Google develop tweaks to its search algorithm, a powerhouse of automation, fielding 100 billion queries a month[.][23]

So even though the algorithms are getting better, they can't do it alone. This insight has led to new, technology-based ways to organize and accomplish work.

In the middle of the last decade, the online retail giant Amazon realized that there were more than a few duplicates among its millions of pages describing products for sale. Algorithms alone didn't do a great job of finding them all, so a team led by employee Peter Cohen built software that showed possible duplicates to human beings and let them make the final determination.[24] Cohen and Amazon soon realized that this was a generally useful innovation. It took a large problem (finding the duplicates among millions of pages), broke it down into many small tasks (are these two pages duplicates?), sent the tasks out to a large group of people, collected their responses, and used them to make progress on the problem (eliminating the duplicates).

The software was originally intended only for internal use, but in November of 2005 Amazon released it to the public under the name Mechanical Turk, in honor of a famous eighteenth-century chess-playing 'robot' that turned out to have a human inside it.[25] The Mechanical Turk software was similar to this automaton in that it too appeared to accomplish tasks automatically, but in reality made use of human labor. It was an example of what Amazon CEO Jeff Bezos called "artificial artificial intelligence," and another way for people to race with machines, although not one with particularly high wages.[26]

Mechanical Turk, which quickly became popular, was an early instance of what came to be called *crowdsourcing*, defined by communications scholar Daren Brabham as "an online, distributed problem-solving and production model."[27] This model is interesting because instead of using technology to automate a process, crowdsourcing makes it deliberately labor intensive. The labor is provided not by a preidentified group of employees, as is the case with most industrial processes, but instead by one or more people (often many more), not identified in advance, who choose to participate.

In less than a decade, crowdsourced production has become an important phenomenon. In fact, it's given rise to a large new crop of companies, often grouped together as the 'peer economy.' Peer economy companies satisfy their customers' requests by crowdsourcing them. Some of the graphs you see in this book, for example, were generated or improved by people we'd never met before. We found them by posting a request for help with the task to TaskRabbit, a company founded by software engineer Leah Busque in 2008. Busque got the idea for TaskRabbit after she ran out of dog food one night and realized that there was no quick and easy way for her to use the Internet to find (and pay) someone willing to pick some up for her.[28]

That same year, Joe Gebbia, Brian Chesky, and Nathan Blecharczyk also launched a website that used the Internet and the crowd to better match supply and demand. In their case, the demand was not for help with a task, but instead for a place to stay. The site, Airbedandbreakfast.com, allowed people to offer rooms in their homes to visitors; it grew out of an experience that Gebbia and Chesky had offering space in their apartment to attendees of a 2007 design conference in San Francisco, where affordable hotel rooms were scarce.

The service they built, which was renamed Airbnb.com in 2009, quickly became popular. On New Year's Eve of 2012, for example, over 140,000 people around the world stayed in places booked via Airbnb; this is 50 percent more than could be accommodated in all the hotels on the Las Vegas Strip.[29] TaskRabbit also grew quickly; by January 2013 the company was reporting "month-over-month transactional growth in the double digits."[30]

TaskRabbit allows people to offer their labor to the crowd while Airbnb lets them offer an asset. The peer economy now includes many examples of both types of company. Crowdsourced labor markets exist in specific domains like programming, design, and cleaning, as well as for general task execution. And people now use websites and apps to rent out their cameras, tools, bicycles, parking spaces, dog kennels, and almost anything else they might own.

Some services bring these two models together and let people offer a combination of labor and assets over the Internet. When Andy needed to have his motorcycle towed to another state in 2010, he found the right person for the job—someone with both time and a trailer on their hands—on uShip. Lyft, founded in 2011, allows people to effectively turn their cars into taxis whenever they want, giving crosstown rides to others. In an effort to avoid opposition from taxi regulators and other authorities, Lyft does not set fees or rates. It instead suggests to customers a 'donation' that they should offer to the person who just gave them a lift.

As the story of Lyft highlights, there are many legal and regulatory issues that will need to be resolved as the peer economy grows. While we certainly acknowledge the need to ensure public safety, we hope that regulation in this new area will not be stifling and that the peer economy will continue to grow. We like the efficiency gains and price declines that crowdsourcing brings, but we also like the work that it brings. Participation in services like TaskRabbit and Airbnb gives people previously unavailable economic opportunities, and it also gives them something to do. It therefore has the potential to address all three of Voltaire's "great evils," and so should be encouraged by policy, regulation, incentives like the ETIC, and other available levers.

The peer economy is still new and still small, both relative to GDP and in absolute terms. In April 2013, for example, TaskRabbit was adding one thousand new people each month to its network of approved task completers.[31] This is encouraging, but that same month there were nearly 4.5 million Americans who had been out of work for at least twenty-seven weeks.[32] Comparisons like this strongly suggest that crowdsourcing does not yet play a significant role in reducing unemployment and bringing work to the economy as a whole.

This fact does not mean that the peer economy should not be encouraged and supported. Quite the opposite. The best solutions—probably, in fact, the only real solutions—to the labor force challenges that will arise in the future will come from markets and capitalism, and from the technology-enabled creations of innovators and entrepreneurs. Peer economy companies are examples of innovations that increase the value of human labor rather than reducing it. Because we believe that work is so important, we believe that policy makers should encourage such creations.

Wild Ideas Welcomed

We've discussed the future and how to shape it with a variety of technologists and labor leaders, with economists and sociologists, with entrepreneurs and retail clerks, and even with science-fiction authors, and we've been impressed with the breadth of ideas offered. This brainstorming is valuable because we are going to need more novel and radical ideas—more 'out-of-the-box thinking'—to deal with the consequences of technological progress. Here are a few of the ideas we've heard. We include them not necessarily to endorse them, but instead to spur further thinking about what kinds of interventions will be effective as machines continue to race ahead.

* Create a national mutual fund distributing the ownership of capital widely and perhaps inalienably, providing a dividend stream to all citizens and assuring the capital returns do not become highly concentrated.
* Use taxes, regulation, contests, grand challenges, or other incentives to try to direct technical change toward machines that augment human ability rather than substitute for it, toward new goods and services and away from labor savings.
* Pay people via nonprofits and other organizations to do 'socially beneficial' tasks, as determined by a democratic process.
* Nurture or celebrate special categories of work to be done by humans only. For instance, care for babies and young children, or perhaps the dying, might fall into this category.
* Start a 'made by humans' labeling movement, similar to those now in place for organic foods, or award credits for companies that employ humans, similar to the carbon offsets that can be purchased. If some consumers wanted to increase the demand for human workers, such labels or credits would let them do so.
* Provide vouchers for basic necessities like food, clothing, and housing, eliminating the extremes of poverty but letting the market manage income above that level.
* Ramp up hiring by the government via programs like the

Depression-era Civilian Conservation Corps to clean up the environment, build infrastructure, and address other public goods. A variant is to increase the role of 'workfare,' i.e., direct payments tied to a work requirement.

Each of these ideas has promising aspects as well as flaws. We don't doubt that there may be other ideas that would be even more effective.*

Of course, theorizing alone has its limits. Perhaps the best advice we can give is to encourage policy experimentation and seek opportunities to systematically test ideas and learn from both successes and failures. In fact, there are individuals, industries, and even whole nations where some aspects of second-machine-age economics are visible today. There are lessons to be learned. For instance: How do lottery winners react to not having to work anymore? (Hint: not always well.) What can we learn from industries with a concentration of high-income superstars like professional sports, motion pictures, and music? What challenges and opportunities do citizens of nations like Norway and the United Arab Emirates face when they have access to enormous wealth as a birthright via sovereign wealth funds? What were the institutions and incentives that helped some children of wealthy landowners in the seventeenth century go on to lead happy, inventive, and creative lives, while others did not?

In the coming decade, we will have the good fortune to witness a wave of astonishing technologies unleashed. They will require changes in our economic institutions and intuitions. By maximizing the flexibility of our systems and mental models, we will be in the best position to identify and implement these changes. A willingness to learn from others' ideas and adapt our own practices—to have open minds and open systems—will be the hallmarks of success.

* We're interested in hearing which ideas you like best, and others you would like to suggest. Contact us at www.SecondMachineAge.com to share your insights.

CHAPTER 15

TECHNOLOGY AND THE FUTURE

(WHICH IS VERY DIFFERENT FROM "TECHNOLOGY *IS* THE FUTURE")

"The machine does not isolate man from the great problems of nature but plunges him more deeply into them."

—Antoine de Saint-Exupery

IT'S ONE OF HUMANITY'S most ancient fantasies: that someday we can all have our material needs fulfilled without drudgery, freeing us to pursue our true interests, amusements, or passions. And that someday, no one will have to toil at an unpleasant task because food, clothing, shelter, and all the other basics for living will be provided by automated servants that do all our bidding. It makes for some great stories. But for most of history, they've been just that: legends and myths populated by fantastical automatons made of clay (like the Jewish golem or Norse giant Mokkerkalfe, built to battle Thor), gold (in the *Iliad*, Homer describes the servants and self-driving tripods built from the precious metal by the god Hephaestus), or leather and wood (the flesh and bone of the artificial man made by craftsman Yanshi in the ancient Chinese Liezi text). The materials change, but the dream remains the same.

To at last make real the dream of human freedom via machine labor, we're using silicon, metal, and plastic. These are the key physical ingredients of the second machine age, at the heart of the digital computers, cables, and sensors being built and deployed with such speed all around the world.

What they're enabling is something without precedent. For all previous generations, when people thought of the best minds of their time working with available materials to make artificial helpers, all they could come up with were stories.

Our generation is different.

Now when we imagine a machine doing a human task, we can be confident that if the automaton doesn't already exist there's at least a good chance that someone in a lab or garage somewhere is tinkering with version 0.1. Over the past three years, the two of us have visited a lot of these innovators and their workshops, and we've been astonished by the brilliant technologies of the second machine age.

After surveying the landscape, we are convinced that we are at an inflection point—the early stages of a shift as profound as that brought on by the Industrial Revolution. Not only are the new technologies exponential, digital, and combinatorial, but most of the gains are still ahead of us. In the next twenty-four months, the planet will add more computer power than it did in all previous history. Over the next twenty-four years, the increase will likely be over a thousand-fold. We've already digitized exabytes of information, but the amount of data that's being digitized is growing even faster than Moore's Law.

Our generation will likely have the good fortune to experience two of the most amazing events in history: the creation of true machine intelligence and the connection of all humans via a common digital network, transforming the planet's economics. Innovators, entrepreneurs, scientists, tinkerers, and many other types of geeks will take advantage of this cornucopia to build technologies that astonish us, delight us, and work for us. Over and over again, they'll show how right Arthur C. Clarke was when he observed that a sufficiently advanced technology can be indistinguishable from magic.

The Risks We'll Run

As we've seen, however, not all the news is good. The middle chapters of this book have shown that while the bounty brought by technology is increasing, so is the spread. And greater spread is not the only possible negative consequence of the coming era of brilliant technology. Our era will face other challenges, ones that are not rooted in economics.

As we move deeper into the second machine age these perils, from both accident and malice, will become greater while material wants and needs are likely to be relatively less important. We will be increasingly concerned with questions about catastrophic events, genuine existential risks, freedom versus tyranny, and other ways that technology can have unintended or unexpected side effects.

The sheer density and complexity of our digital world brings risk with it. Our technological infrastructure is becoming ever more complicated and interlinked. The Internet and intranets, for example, now connect not just people and computers but also televisions, thermostats, burglar alarms, industrial sensors and controls, locomotives, automobiles, and an uncountable multitude of other devices. Many of these provide feedback to one another, and most rely on a few common subsystems like the routers that direct Internet traffic.

Any system this complex and tightly coupled has two related weaknesses. First, it's subject to seeing minor initial flaws cascade via an unpredictable sequence into something much larger and more damaging. Such a cascade, which sociologist Charles Perrow labeled a 'system accident' or 'normal accident,' characterized the 1979 meltdown of the Three Mile Island nuclear plant, the August 2003 electrical blackout that affected forty-five million people throughout the U.S. Northeast, and many other incidents.[1]

Second, complex, tightly coupled systems make tempting targets for spies, criminals, and those who seek to wreak havoc. A recent example here is the Stuxnet computer worm, which may have been incubated in government labs. In 2010 Stuxnet hobbled at least one Iranian nuclear facility by perverting the control systems of its Siemens industrial equipment. The worm entered its target sites and spread through them by jumping harmlessly from PC to PC; when it spotted an opportunity, it crossed over to the Siemens machines and did its damage there.[2]

Until recently, our species did not have the ability to destroy itself. Today it does. What's more, that power will reach the hands of more and more individuals as technologies become both more powerful and cheaper—and thus more ubiquitous. Not all of those individuals will be both sane and well intentioned. As Bill Joy and others have noted, genetic engineering and artificial intelligence can create self-replicating entities.[3] That means that someone working in a basement laboratory might someday use one of these technologies to unleash destructive forces that affect the entire planet. The same scientific breakthroughs in genome sequencing that can be used to cure disease can also be used to create a weaponized version of the smallpox virus.[4] Computer programs can also self-replicate, becoming digital viruses, so the same global network that spreads ideas and innovations can also spread destruction. The physical limits on how much damage any individual or small group could do are becoming less and less constrained. Will our ability to detect and counteract destructive uses of technology advance rapidly enough to keep us safe? That will be an increasingly important question to answer.

George Orwell, William Gibson, and others have described dystopian scenarios involving the loss of freedom and the use of technology to empower despotic rulers and control information flows. Eric Schmidt and Jared Cohen describe some of these technologies, as well as countermeasures, in their book, *The New Digital Age*. The same tools that make it possible to monitor the world in greater detail also give governments and their adversaries the ability to monitor what people are doing and who they are communicating with. There's a genuine tension between our ability to know more and our ability to prevent others from knowing about us. When information was mostly analog and local, the laws of physics created an automatic zone of privacy. In a digital world, privacy requires explicitly designed institutions, incentives, laws, technologies, or norms about which information flows are permitted or prevented and which are encouraged or discouraged.

There are myriad other ways that technology can have unexpected side effects, from addictive gaming and digital distractions to the cyberbalkanization of interest groups, from social isolation to environmental degradation.[5] Even seemingly benevolent inventions, like a technology that dramatically increased longevity, would create enormous social upheaval.*

Is the Singularity Near?

The final, and most far-out, possibility is another sci-fi staple: the development of fully conscious machines. There are two main strands of thinking—one dystopian, one utopian—about what will happen when computers and robots get 'real' minds. The dystopian one finds expression in the *Terminator* and *Matrix* movies and countless other pieces of science fiction. It makes for compelling entertainment, and it seems more and more plausible as technology continues to advance and demonstrate human-like capabilities. Teamwork, after all, is another of these capabilities, so why wouldn't future versions of Watson, the Google autonomous car, the BigDog robot from Boston Dynamics, drone aircraft, and lots of other smart machines decide to work together? And if they did, wouldn't they soon realize that we humans treat our technologies pretty poorly, scrapping them without a second thought? Self-preservation alone would plausibly motivate this digital army to fight against us (perhaps using Siri as an interpreter for the enemy).

In utopian versions of digital consciousness, we humans don't fight with machines; we join with them, uploading our brains into the cloud and otherwise becoming part of a "technological singularity." This is a term coined in 1983 by science-fiction author Vernor Vinge, who predicted that, "We will soon create intelligences greater than our own. . . . When this happens, human history will have reached

* Greg Mankiw ponders a thought experiment where a pill is discovered that adds one year of life to anyone who takes it, but costs $100,000 per pill to produce—more than most people could afford. Would we ban it, ration it, or regulate it in some way?

a kind of singularity, an intellectual transition as impenetrable as the knotted space-time at the center of a black hole, and the world will pass far beyond our understanding."[6]

Progress toward such a singularity, Vinge and others have argued, is driven by Moore's Law. Its accumulated doubling will eventually yield a computer with more processing and storage capacity than the human brain. Once this happens, things become highly unpredictable. Machines could become self-aware, humans and computers could merge seamlessly, or other fundamental transitions could occur. Ray Kurzweil, who has done more than anyone else to explain the power of exponential improvement, wrote in his 2005 book *The Singularity Is Near* that at current rates of progress these transitions will occur by about 2045.[7] How plausible is singularity or the Terminator? We honestly don't know. As with all things digital it's wise never to say never, but we still have a long way to go.

The science-fiction capabilities of *Jeopardy!*-champion supercomputers and autonomous cars can be misleading. Because they're examples of digital technologies doing human-like things, they can lead us to conclude that the technologies themselves are becoming human-like. But they're not—yet. We humans build machines to do things that we see being done in the world by animals and people, but we typically don't build them the same way that nature built us. As AI trailblazer Frederick Jelinek put it beautifully, "Airplanes don't flap their wings."[8]

It's true that scientists, engineers, and other innovators often take cues from biology as they're working, but it would be a mistake to think that this is always the case, or that major recent AI advances have come about because we're getting better at mimicking human thought. Journalist Stephen Baker spent a year with the Watson team to research his book *Final Jeopardy!*. He found that, "The IBM team paid little attention to the human brain while programming Watson. Any parallels to the brain are superficial, and only the result of chance."[9]

As we were researching this book we heard similar sentiments from most of the innovators we talked to. Most of them weren't trying to unravel the mysteries of human consciousness or understand exactly how we think; they were trying to solve problems and seize opportunities. As they did so, they sometimes came up with technologies that had human-like skills and abilities. But these tools themselves were not like humans at all. Current AI, in short, looks intelligent, but it's an artificial resemblance. That might change in the future. We might start to build digital tools that more closely mimic our minds, perhaps even drawing on our rapidly improving capabilities for scanning and mapping brains. And if we do, those digital minds will certainly augment ours and might even eventually merge with them, or become self-aware on their own.

Destined For . . . ?

Even in the face of all these challenges—economic, infrastructural, biological, societal, and existential—we're still optimistic. To paraphrase Martin Luther King, Jr., the arc of history is long but it bends towards justice.[10] We think the data support this. We've seen not just vast increases in wealth but also, on the whole, more freedom, more social justice, less violence, and less harsh conditions for the least fortunate and greater opportunities for more and more people.

In Charles Dickens's *A Christmas Carol*, when the Ghost of Christmas Future pointed at Scrooge's tombstone Scrooge asked, "Is this what must be, or what might be?" For questions of technology and the future state of the world, it's the latter. Technology creates possibilities and potential, but ultimately, the future we get will depend on the choices we make. We can reap unprecedented bounty and freedom, or greater disaster than humanity has ever seen before.

The technologies we are creating provide vastly more power to change the world, but with that power comes greater responsibility. That's why we aren't technological determinists, and that's why we devoted three chapters in this book to a set of recommendations that we think will improve the odds of achieving a society with shared prosperity.

But in the long run, the real questions will go beyond economic growth. As more and more work is done by machines, people can spend more time on other activities. Not just leisure and amusements, but also the deeper satisfactions that come from invention and exploration, from creativity and building, and from love, friendship, and community. We don't have a lot of formal metrics for those kinds of value, and perhaps we never will, but they will nonetheless grow in importance as we satisfy our more basic economic needs. If the first machine age helped unlock the forces of energy trapped in chemical bonds to reshape the physical world, the real promise of the second machine age is to help unleash the power of human ingenuity.

Our success will depend not just on our technological choices, or even on the coinvention of new organizations and institutions. As we have fewer constraints on what we can do, it is then inevitable that our *values* will matter more than ever. Will we choose to have information widely disseminated or tightly controlled? Will our prosperity be broadly shared? What will be the nature and magnitude of rewards we give to our innovators? Will we build vibrant relationships and communities? Will everyone have the opportunities to discover, create, and enjoy the best of life?

In the second machine age, we need to think much more deeply about what it is we really want and what we value, both as individuals and as a society. Our generation has inherited more opportunities to transform the world than any other. That's a cause for optimism, but only if we're mindful of our choices.

Technology is not destiny. We shape our destiny.

ACKNOWLEDGMENTS

There's a general story of how this book came to be, and a specific one. Many people contributed to each, and some to both.

The general story is about our research to understand the nature of progress with digital technologies, and its economic and societal consequences. As part of this work, we talked to two main types of geek (a label which, to us, is the highest praise): those who study economics and other social sciences, and those who build technologies. In the former group Susan Athey, David Autor, Zoe Baird, Nick Bloom, Tyler Cowen, Charles Fadel, Chrystia Freeland, Robert Gordon, Tom Kalil, Larry Katz, Tom Kochan, Frank Levy, James Manyika, Richard Murnane, Robert Putnam, Paul Romer, Scott Stern, Larry Summers, and Hal Varian have helped our thinking enormously. In the latter category are Chris Anderson, Rod Brooks, Peter Diamandis, Ephraim Heller, Reid Hoffman, Jeremy Howard, Kevin Kelly, Ray Kurzweil, John Leonard, Tod Loofbourrow, Hilary Mason, Tim O'Reilly, Sandy Pentland, Brad Templeton, and Vivek Wadhwa. All of them were incredibly generous with their time and tolerant of our questions. We did our best to understand the insights they shared with us, and apologize for whatever mistakes we made in trying to convey them in this book.

Some members of both groups came together at an extraordinary series of lunches at MIT organized by John Leonard, Frank

Levy, Daniela Rus, and Seth Teller that assembled people from the Economics Department, the Sloan School of Management, and the Computer Science and Artificial Intelligence Lab to talk about exactly the topics in which we were most interested. We had truly cross-disciplinary conversations without any forcing function other than our own curiosities, which have remained strong enough to resist the countless hectoring demands of academic life.

As these lunches indicate, MIT itself is part of the general story of this book. It's been the ideal professional home for us, and we're grateful for the support we've received from Sloan, its dean David Schmittlein, and deputy dean S. P. Kothari. The intellects at MIT make it a humbling place; the people make it a lovely one.

The specific story of this book starts with an inquiry we got from Raphael Sagalyn who, we soon learned, is a lion among literary agents (he was introduced to us by Joan Powell, Andy's equally illustrious speaking agent). Rafe wanted to know if we had any interest in turning our short, self-published e-book *Race Against the Machine* into a real book, one with a publisher, an editor, a hard cover—the works. Rafe was far too professional to use the word "real," of course, but we knew what he meant.

And we were intrigued, because we hadn't stopped thinking and talking to each other about the ideas in *Race Against the Machine* even after the book was done. In fact, we'd only become more interested in the concepts of technological progress and its economic consequences as a result of the e-book, and we'd enjoyed the many conversations it had sparked with people all over the world. So it didn't take long at all for us to decide to work with Rafe to see if this interest was shared by mainstream publishers.

Amazingly enough, it was, which is how we met our editor, Brendan Curry, and his colleagues at W. W. Norton. Working under tight deadlines, Brendan and his colleagues Mitchell Kohles and Tara Powers guided our manuscript into shape. We're grateful for their advice and keen attention, which were delivered with grace under pressure.

At the intersection where our general interests met the specifics of writing a book is a set of colleagues, family, and friends who we simply can't thank enough. To give us up-close encounters with the technologies we were writing about, Dave Ferrucci and his colleagues at IBM brought Watson to campus, Rod Brooks introduced us to Baxter the humanoid robot, Carl Bass at Autodesk headquarters let us handle a range of objects made by 3D printing, and Betsy Masiello and Hal Varian worked their magic at Google to get us a ride in one of their driverless cars. We're grateful for the students in our classes who served as sounding boards for many of the ideas that made it into this book, and even more that didn't make the cut.

We are particularly grateful to our Digital Frontier team, a self-selecting group of people who are interested in the same things we are, and who get together periodically to generate, share, and refine ideas, a lot of which made their way into this book. Matt Beane, Greg Gimpel, Shan Huang, Heekyung Kim, Tod Loofbourrow, Frank MacCrory, Max Novendstern, JooHee Oh, Shachar Reichman, Guillaume Saint Jacques, Michael Schrage, Dipak Shetty, Gabriel Unger, and George Westerman helped us explore the digital frontier. Matt and Dipak went above and beyond the call by helping us with many of the graphs that appear here, as did Gabriel, George, Greg, Michael, and Tod by giving us detailed comments on the manuscript. Max put in countless hours under tight deadline checking facts. Meghan Hennessey managed Erik's increasingly crowded work schedule, while Martha Pavlakis's strength, courage, and grace as she battled and defeated cancer reminded him what really matters in life. Esther Simmons kept Andy on track and on time, his family kept him sane, and Tatiana Lingos-Webb constantly gave him reasons to smile (which is no small task at times).

Finally, our colleagues at the MIT Center for Digital Business and Initiative on the Digital Economy deserve more thanks than we can articulate. Tammy Buzzell and Justin Lockenwitz keep the place running like clockwork, and executive director David Verrill continues to astound us with everything he does and how easy he makes it all look. We've said it before, but it bears repeating: no matter what skills and abilities technology ever acquires, it won't come anywhere near him.

NOTES

Chapter 1 THE BIG STORIES

1. Ian Morris, *Why the West Rules—For Now: The Patterns of History, and What They Reveal About the Future* (New York: Farrar, Straus and Giroux, 2010), p. 73.
2. Ibid., p. 74.
3. Ibid., p. 71.
4. Ibid., p. 112.
5. Karl Jaspers, *The Origin and Goal of History. Translated From the German by Michael Bullock* (London: Routledge K. Paul, 1953), p. 51.
6. "Major Religions of the World Ranked by Number of Adherents," 2007, http://www.adherents.com/Religions_By_Adherents.html/.
7. Anne Rooney, *The History of Mathematics* (New York: The Rosen Publishing Group, 2012), p. 18.
8. Morris, *Why the West Rules—For Now*, p. 142.
9. Louis C. Hunter and Eleutherian Mills-Hagley Foundation, *A History of Industrial Power in the United States, 1780–1930: Steam Power* (Charlottesville, VA: University Press of Virginia, 1979), 601–30.
10. Morris, *Why the West Rules—For Now*, p. 497.
11. Ibid., p. 492.
12. Martin L. Weitzman, "Recombinant Growth," *Quarterly Journal of Economics* 113, no. 2 (1998): 331–60.
13. Bjørn Lomborg, *The Skeptical Environmentalist: Measuring the Real State of the World* (Cambridge, UK: Cambridge University Press, 2001), p. 165.

Chapter 2 THE SKILLS OF THE NEW MACHINES

1. Frank Levy and Richard J. Murnane, *The New Division of Labor: How Computers Are Creating the Next Job Market* (Princeton, NJ: Princeton University Press, 2004).
2. Michael Polanyi, *The Tacit Dimension* (Chicago, IL: University of Chicago Press, 2009), p. 4.

3. Joseph Hooper, "DARPA's Debacle in the Desert," *Popular Science*, June 4, 2004, http://www.popsci.com/scitech/article/2004-06/darpa-grand-challenge-2004 darpas-debacle-desert.
4. Mary Beth Griggs, "4 Questions About Google's Self-Driving Car Crash," *Popular Mechanics*, August 11, 2011, http://www.popularmechanics.com/cars/news/indus try/4-questions-about-googles-self-driving-car-crash; John Markoff, "Google Cars Drive Themselves, in Traffic," *New York Times*, October 9, 2010, http://www.nytimes .com/2010/10/10/science/10google.html.
5. Ernest Hemingway, *The Sun Also Rises* (New York: HarperCollins, 2012), p. 72.
6. Levy and Murnane, *The New Division of Labor*, p. 29.
7. "Siri Is Actually Incredibly Useful Now," *Gizmodo*, accessed August 4, 2013, http://gizmodo.com/5917461/siri-is-better-now.
8. Ibid.
9. "Minneapolis Street Test: Google Gets a B+, Apple's Siri Gets a D - Apple 2.0 -Fortune Tech," *CNNMoney*, http://tech.fortune.cnn.com/2012/06/29/minneapolis-street-test-google-gets-a-b-apples-siri-gets-a-d/ (accessed June 23, 2013).
10. Ning Xiang and Rendell Torres, "Architectural Acoustics and Signal Processing in Acoustics: Topical Meeting on Spatial and Binaural Evaluation of Performing Arts Spaces I: Measurement Techniques and Binaural and Interaural Modeling," 2004, http://scita tion.aip.org/getpdf/servlet/GetPDFServlet?filetype=pdf&id=JASMAN000116000004.
11. As quoted in John Markoff, "Armies of Expensive Lawyers, Replaced by Cheaper Software," *New York Times*, March 4, 2011, http://www.nytimes.com/2011/03/05/science/05legal.html?pagewanted=all&_r=0.
12. "Spring Cleaning for Some of Our APIs," *The Official Google Code Blog*, June 3, 2011, http://googlecode.blogspot.com/2011/05/spring-cleaning-for-some-of-our-apis .html.
13. Douglas Adams, *The Hitchhiker's Guide to the Galaxy* (New York: Random House, 2007), p. 54.
14. Personal communication with Sara Buda, Lionbridge Vice President, Investor Relations and Corporate Development, September 2011.
15. "Top 10 TV Ratings / Top 10 TV Shows / Nielsen," *Evernote*, August 18, 2012, https://www.evernote.com/shard/s13/sh/a4480367-9414-4246-bba4-d588d60e64ce/bb3f 380315cd10deef79e33a88e56602 (accessed June 23, 2013).
16. "Meet Watson, the Jeopardy!-Playing Computer," *TV.com*, December 1, 2004, http://www.tv.com/news/meet-watson-the-jeopardy-playing-computer-25144/.
17. "What's The Most Money Won On Jeopardy?," *Celebrity Net Worth*, May 20, 2010, http://www.celebritynetworth.com/articles/entertainment-articles/whats-the-most-money-won-o/.
18. Stephen Baker, *Final Jeopardy: Man Vs. Machine and the Quest to Know Everything* (Houghton Mifflin Harcourt, 2011), p. 19.
19. "IBM and 'Jeopardy!' Relive History With Encore Presentation of 'Jeopardy!'," *Did You Know . . .*, 2013, http://www.jeopardy.com/showguide/abouttheshow/showhistory/.

20. All Watson and human performance statistics from Willy Shih, "Building Watson: Not So Elementary, My Dear!" Harvard Business School Case 612-017, September 2011 (revised July 2012), http://hbr.org/product/building-watson-not-so-elementary-my-dear/an/612017-PDF-ENG.

21. Authors' personal research.

22. Ken Jennings, "My Puny Human Brain," *Slate*, February 16, 2011, http://www.slate.com/articles/arts/culturebox/2011/02/my_puny_human_brain.single.html.

23. Isaac Asimov, "The Vocabulary of Science Fiction," in *Asimov on Science Fiction* (New York, Doubleday, 1981), p. 69.

24. "The Robot Panic of the Great Depression," *Slate*, November 29, 2011, http://www.slate.com/slideshows/technology/the-robot-panic-of-the-great-depression.html (accessed June 23, 2013).

25. "Isaac Asimov Explains His Three Laws of Robots," *Open Culture*, October 31, 2012, http://www.openculture.com/2012/10/isaac_asimov_explains_his_three_laws_of_robotics.html (accessed June 23, 2013).

26. Brian Lam, "Honda ASIMO vs. Slippery Stairs," December 11, 2006, http://gizmodo.com/220771/honda-asimo-vs-slippery-stairs?op=showcustomobject&postId=220771&item=0.

27. Hans Moravec, *Mind Children: The Future of Robot and Human Intelligence* (Cambridge, MA: Harvard University Press, 1988), p. 15.

28. "Moravec's Paradox," *Wikipedia, the Free Encyclopedia*, April 28, 2013, http://en.wikipedia.org/w/index.php?title=Moravecpercent27s_paradox&oldid=540679203.

29. Steven Pinker, *The Language Instinct* (New York: HarperPerennial ModernClassics, 2007), p. 190–91.

30. Christopher Drew, "For iRobot, the Future Is Getting Closer," *New York Times*, March 2, 2012, http://www.nytimes.com/2012/03/03/technology/for-irobot-the-future-is-getting-closer.html.

31. Danielle Kucera, "Amazon Acquires Kiva Systems in Second-Biggest Takeover," *Bloomberg*, March 19, 2012, http://www.bloomberg.com/news/2012-03-19/amazon-acquires-kiva-systems-in-second-biggest-takeover.html (accessed June 23, 2013).

32. Marc DeVidts, "First Production Run of Double Has Sold Out!," August 16, 2012, http://blog.doublerobotics.com/2012/8/16/welcome-double-update.

33. "DARPA Robotics Challenge," n.d., http://www.darpa.mil/Our_Work/TTO/Programs/DARPA_Robotics_Challenge.aspx.

34. DARPA, "Broad Agency Announcement DARPA Robots Challenge Tactical Technology Office," April 10, 2012, http://www.fbo.gov/utils/view?id=74d674ab011d5954c7a46b9c21597f30.

35. For instance, *Philips Vital Signs Camera*, n.d., http://www.vitalsignscamera.com/; Steve Casimiro, "2011 Best Outdoor iPhone Apps—Best Weather Apps," n.d., http://www.adventure-journal.com/2011-best-outdoor-iphone-apps-%E2%80%94-best-weather-apps/; *iSeismometer*, n.d., https://itunes.apple.com/us/app/iseismometer/id304190739?mt=8.

36. "SoLoMo," *Schott's Vocab Blog*, http://schott.blogs.nytimes.com/2011/02/22/solomo/ (accessed June 23, 2013).
37. "SCIgen – An Automatic CS Paper Generator," accessed September 14, 2013, http://pdos.csail.mit.edu/scigen/.
38 Herbert Schlangemann, "Towards the Simulation of E-commerce," in *Proceedings of the 2008 International Conference on Computer Science and Software Engineering*, vol. 5, CSSE 2008 (Washington, D.C.: IEEE Computer Society, 2008), 1144–47, doi:10.1109/CSSE.2008.1.
39. Narrative Science, "Forbes Earnings Preview: H.J. Heinz," August 24, 2012, http://www.forbes.com/sites/narrativescience/2012/08/24/forbes-earnings-preview-h-j-heinz-3/.
40. "How Stereolithography 3-D Layering Works," *HowStuffWorks*, http://computer.howstuffworks.com/stereolith.htm (accessed August 4, 2013).
41. Claudine Zap, "3D Printer Could Build a House in 20 Hours," August 10, 2012, http://news.yahoo.com/blogs/sideshow/3d-printer-could-build-house-20-hours-224156687.html; see also Samantha Murphy, "Woman Gets Jawbone Made By 3D Printer," February 6, 2012, http://mashable.com/2012/02/06/3d-printer-jawbone/; "Great Ideas Soar Even Higher with 3D Printing," 2013, http://www.stratasys.com/resources/case-studies/aerospace/nasa-mars-rover.

Chapter 3 MOORE'S LAW AND THE SECOND HALF OF THE CHESSBOARD

1. G. E. Moore, "Cramming More Components onto Integrated Circuits," *Electronics* 38, no. 8 (April 19, 1965): 114–17, doi:10.1109/jproc.1998.658762.
2. Ibid.
3. Michael Kanellos, "Moore's Law to Roll on for Another Decade," *CNET*, http://news.cnet.com/2100-1001-984051.html (accessed June 26, 2013).
4. Rick Merritt, "Broadcom: Time to Prepare for the End of Moore's Law," *EE Times*, May 23, 2013, http://www.eetimes.com/document.asp?doc_id=1263256.
5. Adam Sneed, "A Brief History of Warnings About the Demise of Moore's Law," *Future Tense* blog, Slate.com, May 3, 2012, http://www.slate.com/blogs/future_tense/2012/05/03/michio_kako_and_a_brief_history_of_warnings_about_the_end_of_moore_s_law.html (accessed June 26, 2013).
6. "Moore's Law: The Rule That Really Matters in Tech," *CNET*, October 15, 2012, http://news.cnet.com/8301-11386_3-57526581-76/moores-law-the-rule-that-really-matters-in-tech/.
7. H. J. R Murray, *A History of Chess* (Northampton, MA: Benjamin Press, 1985).
8. Ray Kurzweil, *The Age of Spiritual Machines: When Computers Exceed Human Intelligence* (London: Penguin, 2000), p. 36.
9. See http://www.cuug.ab.ca/~branderr/pmc/012_coal.html (accessed September 23, 2013).

10. Ionut Arghire, "The Petaflop Barrier Is Down, Going for the Exaflop?," *Softpedia*, June 10, 2008, http://news.softpedia.com/news/The-Petaflop-Barrier-Is-Down-Going-for-the-Exaflop-87688.shtml.
11. "The Tops in Flops," *Scribd*, http://www.scribd.com/doc/88630700/The-Tops-in-Flops (accessed June 26, 2013).
12. Matt Gemmell, "iPad Multi-Touch," May 9, 2010, http://mattgemmell.com/2010/05/09/ipad-multi-touch/.
13. "Company News; Cray to Introduce A Supercomputer," *New York Times*, February 11, 1988, http://www.nytimes.com/1988/11/02/business/company-news-cray-to-introduce-a-supercomputer.html (accessed June 26, 2013).
14. Thomas Fine, "The Dawn of Commercial Digital Recording," *ARSC Journal* 39 (Spring 2008): 1–17; Jurrien Raif, "Steven Sasson Named to CE Hall of Fame," *Let's Go Digital*, September 18, 2007, http://www.letsgodigital.org/en/16859/ce-hall-of-fame/.
15. "Hendy's Law," Nida Javed, December 7, 2012, http://prezi.com/v-rooknipogx/hendys-law/.
16. Josep Aulinas et al., "The SLAM Problem: A Survey," in *Proceedings of the 2008 Conference on Artificial Intelligence Research and Development: Proceedings of the 11th International Conference of the Catalan Association for Artificial Intelligence* (Amsterdam: IOS Press, 2008), pp. 363–71, http://dl.acm.org/citation.cfm?id=1566899.1566949.
17. Dylan McGrath, "Teardown: Kinect Has Processor after All," *EE Times*, November 15, 2010, http://www.eetimes.com/electronics-news/4210757/Teardown—Kinect-has-processor-after-all.
18. "Microsoft Kinect Sales Top 10 Million, Set New Guinness World Record," *Mashable*, March 9, 2011, http://mashable.com/2011/03/09/kinect-10-million/ (accessed June 26, 2013).
19. "Xbox Kinect's Game Launch Lineup Revealed," *Mashable*, October 18, 2010, http://mashable.com/2010/10/18/kinect-launch-games/ (accessed June 26, 2013).
20. "KinectFusion: The Self-Hack That Could Change Everything," *The Creators Project*, August 18, 2011, http://thecreatorsproject.vice.com/blog/kinectfusion-the-self-hack-that-could-change-everything (accessed June 26, 2013).
21. Sarah Kessler, "KinectFusion HQ – Microsoft Research," http://research.microsoft.com/apps/video/dl.aspx?id=152815 (accessed June 26, 2013).
22. "Microsoft's KinectFusion Research Project Offers Real-time 3D Reconstruction, Wild AR Possibilities," *Engadget*, August 9, 2011, http://www.engadget.com/2011/08/09/microsofts-kinectfusion-research-project-offers-real-time-3d-re/ (accessed June 26, 2013).
23. Thomas Whelan et al., "Kintinuous: Spatially Extended KinectFusion," n.d., http://dspace.mit.edu/bitstream/handle/1721.1/71756/MIT-CSAIL-TR-2012-020.pdf?sequence=1.
24. Brett Solomon, "Velodyne Creating Sensors for China Autonomous Vehicle Market," *Technology Tell*, July 5, 2013, http://www.technologytell.com/in-car-tech/4283/velodyne-creating-sensors-for-china-autonomous-vehicle-market/.

Chapter 4 THE DIGITIZATION OF JUST ABOUT EVERYTHING

1. Nick Wingfield and Brian X. Chen, "Apple Keeps Loyalty of Mobile App Developers," *New York Times*, June 10, 2012, http://www.nytimes.com/2012/06/11/technology/apple-keeps-loyalty-of-mobile-app-developers.html.
2. "How Was the Idea for Waze Created?," http://www.waze.com/faq/ (accessed June 27, 2013).
3. Daniel Feldman, "Waze Hits 20 Million Users!," July 5, 2012, http://www.waze.com/blog/waze-hits-20-million-users/.
4. Carl Shapiro and Hal R. Varian, *Information Rules: A Strategic Guide to the Network Economy* (Boston, MA: Harvard Business School Press, 1998), p. 3.
5. Jules Verne, *Works of Jules Verne* (New York: V. Parke, 1911), http://archive.org/details/worksofjulesvern01vernuoft.
6. Shapiro and Varian, *Information Rules*, p. 21.
7. "Friendster," *Wikipedia*, http://en.wikipedia.org/w/index.php?title=Friendster&oldid=559301831 (accessed June 27, 2013); "History of Wikipedia," *Wikipedia*, http://en.wikipedia.org/w/index.php?title=History_of_Wikipedia&oldid=561664870 (accessed June 27, 2013); "Blogger (service)," *Wikipedia*, http://en.wikipedia.org/w/index.php?title=Blogger_(service)&oldid=560541931 (accessed June 27, 2013).
8. "Top Sites," *Alexa: The Web Information Company*, http://www.alexa.com/topsites (accessed September 8, 2012).
9. "IBM Watson Vanquishes Human Jeopardy Foes," *PCWorld*, February 16, 2011, http://www.pcworld.com/article/219893/ibm_watson_vanquishes_human_jeopardy_foes.html.
10. "IBM's Watson Memorized the Entire 'Urban Dictionary,' Then His Overlords Had to Delete It," *The Atlantic*, January 10, 2013, http://www.theatlantic.com/technology/archive/2013/01/ibms-watson-memorized-the-entire-urban-dictionary-then-his-overlords-had-to-delete-it/267047/.
11. Kevin J. O'Brien, "Talk to Me, One Machine Said to the Other," *New York Times*, July 29, 2012, http://www.nytimes.com/2012/07/30/technology/talk-to-me-one-machine-said-to-the-other.html.
12. "VNI Forecast Highlights," *Cisco*, http://www.cisco.com/web/solutions/sp/vni/vni_forecast_highlights/index.html (accessed June 28, 2013).
13. "VNI Forecast Highlights," *Cisco*, http://www.cisco.com/web/solutions/sp/vni/vni_forecast_highlights/index.html (accessed June 28, 2013).
14. Infographic, "The Dawn of the Zettabyte Era," *Cisco Blogs*, http://blogs.cisco.com/news/the-dawn-of-the-zettabyte-era-infographic/ (accessed June 28, 2013).
15. Russ Rowlett, "How Many? A Dictionary of Units of Measurement," April 16, 2005, http://www.unc.edu/~rowlett/units/prefixes.html.
16. Rumi Chunara, Jason R. Andrews, and John S. Brownstein, "Social and News Media Enable Estimation of Epidemiological Patterns Early in the 2010 Haitian Cholera Outbreak," *American Journal of Tropical Medicine and Hygiene* 86, no. 1 (2012): 39–45, doi:10.4269/ajtmh.2012.11-0597.

17. Sitaram Asur and Bernardo A. Huberman, *Predicting the Future with Social Media*," arXiv e-print, Cornell University Library, March 29, 2010, http://arxiv.org/abs/1003.5699.
18. Jennifer Howard, "Google Begins to Scale Back Its Scanning of Books From University Libraries," *Chronicle of Higher Education*, March 9, 2012, http://chronicle.com/article/Google-Begins-to-Scale-Back/131109/.
19. "Culturomics," http://www.culturomics.org/ (accessed June 28, 2013).
20. Jean-Baptiste Michel et al., "Quantitative Analysis of Culture Using Millions of Digitized Books," *Science* 331, no. 6014 (2011): 176–82, doi:10.1126/science.1199644.
21. Steve Lohr, "For Today's Graduate, Just One Word: Statistics," *New York Times*, August 6, 2009, http://www.nytimes.com/2009/08/06/technology/06stats.html.
22. Boyan Brodaric, *Field Data Capture and Manipulation Using GSC Fieldlog V3.0*, U.S. Geological Survey Open-File Report 97-269 (Geological Survey of Canada, October 7, 1997), http://pubs.usgs.gov/of/1997/of97-269/brodaric.html.
23. *Selective Availability* (National Coordination Office for Space-Based Positioning, Navigation, and Timing, February 17, 2012), http://www.gps.gov/systems/gps/modernization/sa/.

Chapter 5 INNOVATION: DECLINING OR RECOMBINING?

1. Henry Southgate, *Many Thoughts of Many Minds: Being a Treasury of Reference Consisting of Selections from the Writings of the Most Celebrated Authors . . .* (Griffin, Bohn, and Company, 1862), p. 451.
2. Paul R. Krugman, *The Age of Diminished Expectations: U.S. Economic Policy in the 1990s* (Cambridge, MA: MIT Press, 1997), p. 11.
3. Joseph Alois Schumpeter, *Business Cycles: A Theoretical, Historical, and Statistical Analysis of the Capitalist Process* (Philadelphia, NJ: Porcupine Press, 1982), p. 86.
4. Robert J. Gordon, *Is U.S. Economic Growth Over? Faltering Innovation Confronts the Six Headwinds*, Working Paper (National Bureau of Economic Research, August 2012), http://www.nber.org/papers/w18315.
5. Ibid.
6. Tyler Cowen, *The Great Stagnation: How America Ate All the Low-hanging Fruit of Modern History, Got Sick, and Will (Eventually) Feel Better* (New York: Dutton, 2011).
7. Gavin Wright, "Review of Helpman (1998)," *Journal of Economic Literature* 38 (March 2000): 161–62.
8. Boyan Jovanovic and Peter L. Rousseau, "General Purpose Technologies," in *Handbook of Economic Growth*, ed. Philippe Aghion and Steven N. Durlauf, vol. 1, Part B (Amsterdam: Elsevier, 2005), 1181–1224, http://www.sciencedirect.com/science/article/pii/S157406840501018X.
9. Alexander J. Field, *Does Economic History Need GPTs?* (Rochester, NY: Social Science Research Network, 2008), http://papers.ssrn.com/abstract=1275023.
10. Gordon, *Is U.S. Economic Growth Over?*, p. 11.
11. Cowen, *The Great Stagnation*, location 520.

12. Gordon, *Is U.S. Economic Growth Over?*, p. 2.
13. Kary Mullis, "The Polymerase Chain Reaction" (Nobel Lecture, December 8, 1993), http://www.nobelprize.org/nobel_prizes/chemistry/laureates/1993/mullis-lecture.html?print=1.
14. W. Brian Arthur, *The Nature of Technology: What It Is and How It Evolves* (New York: Simon and Schuster, 2009), p. 122.
15. Paul Romer, "Economic Growth," *Library of Economics and Liberty*, 2008, http://www.econlib.org/library/Enc/EconomicGrowth.html.
16. Ibid.
17. Associated Press, "Number of Active Users at Facebook over the Years," *Yahoo! Finance*, http://finance.yahoo.com/news/number-active-users-facebook-over-years-214600186—finance.html (accessed June 29, 2013).
18. Martin L. Weitzman, "Recombinant Growth," *Quarterly Journal of Economics* 113, no. 2 (1998): 331–60.
19. Ibid., 357.
20. Eric Raymond, "The Cathedral and the Bazaar," September 11, 2000, http://www.catb.org/esr/writings/homesteading/cathedral-bazaar/.
21. "NASA Announces Winners of Space Life Sciences Open Innovation Competition," *NASA – Johnson Space Center – Johnson News*, http://www.nasa.gov/centers/johnson/news/releases/2010/J10-017.html (accessed June 29, 2013).
22. Steven Domeck, "NASA Challenge Pavilion Results," 2011, http://www.nasa.gov/pdf/651444main_InnoCentive%20NASA%20Challenge%20Results%20CoECI_D1_0915%20to%200955.pdf.
23. Lars Bo Jeppesen and Karim Lahkani, "Marginality and Problem Solving Effectiveness in Broadcast Search," *Organization Science* 20 (2013), http://dash.harvard.edu/bitstream/handle/1/3351241/Jeppesen_Marginality.pdf?sequence=2.
24. "Predicting Liability for Injury from Car Accidents," *Kaggle*, 2013, http://www.kaggle.com/solutions/casestudies/allstate.
25. "Carlsberg Brewery Harnesses Design Innovation Using Affinnova," *Affinnova*, http://www.affinnova.com/success-story/carlsberg-breweries/ (accessed August 6, 2013).

Chapter 6 ARTIFICIAL AND HUMAN INTELLIGENCE IN THE SECOND MACHINE AGE

1. John Markoff, "Israeli Start-Up Gives Visually Impaired a Way to Read," *New York Times*, June 3, 2013, http://www.nytimes.com/2013/06/04/science/israeli-start-up-gives-visually-impaired-a-way-to-read.html.
2. "Press Announcements – FDA Approves First Retinal Implant for Adults with Rare Genetic Eye Disease," *WebContent*, February 14, 2013, http://www.fda.gov/NewsEvents/Newsroom/PressAnnouncements/ucm339824.htm.
3. "Wheelchair Makes the Most of Brain Control," *MIT Technology Review*, Septem-

ber 13, 2010, http://www.technologyreview.com/news/420756/wheelchair-makes-the-most-of-brain-control/.

4. "IBM Watson Helps Fight Cancer With Evidence-based Diagnosis and Treatment Suggestions," *Memorial Sloan-Kettering Cancer Center*, January 2013, http://www-03.ibm.com/innovation/us/watson/pdf/MSK_Case_Study_IMC14794.pdf.

5. David L. Rimm, "C-Path: A Watson-Like Visit to the Pathology Lab," *Science Translational Medicine* 3, no. 108 (2011): 108fs8–108fs8.

6. Andrew H. Beck et al., "Systematic Analysis of Breast Cancer Morphology Uncovers Stromal Features Associated with Survival," *Science Translational Medicine* 3, no. 108 (2011): 108ra113–108ra113, doi:10.1126/scitranslmed.3002564.

7. Julian Lincoln Simon, *The Ultimate Resource* (Princeton, NJ: Princeton University Press, 1981), p. 196.

8. Julian Lincoln Simon, *The Ultimate Resource* 2 (rev. ed., Princeton, NJ: Princeton University Press, 1998), p. xxxviii.

9. World Bank, *Information and Communications for Development 2012: Maximizing Mobile* (Washington, DC: World Bank Publications, 2012).

10. Robert Jensen, "The Digital Provide: Information (Technology), Market Performance, and Welfare in the South Indian Fisheries Sector," *Quarterly Journal of Economics* 122, no. 3 (2007): 879–924, doi:10.1162/qjec.122.3.879.

11. Erica Kochi, "How The Future of Mobile Lies in the Developing World," *TechCrunch*, May 27, 2012, http://techcrunch.com/2012/05/27/mobile-developing-world/.

12. Marguerite Reardon, "Smartphones to Outsell Feature Phones in 2013 for First Time," *CNET*, March 4, 2013, http://news.cnet.com/8301-1035_3-57572349-94/smartphones-to-outsell-feature-phones-in-2013-for-first-time/.

13. Jonathan Rosenblatt, "Analyzing Your Data on the AWS Cloud (with R)," *R-statistics Blog*, July 22, 2013, http://www.r-statistics.com/2013/07/analyzing-your-data-on-the-aws-cloud-with-r/.

14. Carl Bass, "We've Reached Infinity—So Start Creating," *Wired UK*, February 22, 2012, http://www.wired.co.uk/magazine/archive/2012/03/ideas-bank/weve-reached-infinity.

15. Noam Cohen, "Surviving Without Newspapers," *New York Times*, June 7, 2009, http://www.nytimes.com/2009/06/07/weekinreview/07cohen.html.

Chapter 7 COMPUTING BOUNTY

1. While the rate has fluctuated with recession, over longer periods it has been remarkably steady. In fact, in 1957, the economist Nicholas Kaldor summarized what was known about economic growth at the time in a classic article: "A Model of Economic Growth," *Economic Journal* 67, no. 268 (1957): 591–624. His observations, including the relatively constant growth rates of key variables, such as wage growth and the amount of capital per worker, came to be known as the "Kaldor Facts."

2. Bret Swanson,"Technology and the Growth Imperative," *The American*, March 26, 2012, http://www.american.com/archive/2012/march/technology-and-the-growth-imperative (accessed Sept 23, 2013).
3. Congressional Budget Office, *The 2013 Long-Term Budget Outlook*, September 2013, p. 95. http://www.cbo.gov/sites/default/files/cbofiles/attachments/44521-LTBO2013.pdf.
4. Robert Solow, "We'd Better Watch Out," *New York Times Book Review*, July 12, 1987.
5. Erik Brynjolfsson, "The Productivity Paradox of Information Technology," *Communications of the ACM* 36, no. 12 (1993): 66–77, doi:10.1145/163298.163309.
6. See, e.g., Erik Brynjolfsson and Lorin Hitt, "Paradox Lost: Firm Level Evidence on the Returns to Information Systems," *Management Science* 42, no. 4 (1996): 541–58. See also Brynjolfsson and Hitt, "Beyond Computation: Information Technology, Organizational Transformation and Business Performance," *Journal of Economic Perspectives* 14, no. 4 (2000): 23–48, which summarizes much of the literature on this question.
7. Dale W. Jorgenson, Mun S. Ho, and Kevin J. Stiroh, "Will the U.S. Productivity Resurgence Continue?," *Current Issues in Economics and Finance* (2004), http://ideas.repec.org/a/fip/fednci/y2004idecnv.10no.13.html.
8. C. Syverson, "Will History Repeat Itself? Comments on 'Is the Information Technology Revolution Over?' " *International Productivity Monitor 25* (2013): 37–40.
9. "Computer and Dynamo: The Modern Productivity Paradox in a Not-Too-Distant Mirror," *Center for Economic Policy Research*, no. 172, Stanford University, July 1989, http://www.dklevine.com/archive/refs4115.pdf.
10. For instance, Materials Resource Planning (MRP) systems, which begat Enterprise Resource Planning (ERP), and then Supply Chain Management (SCM), Customer Relationship Management (CRM), and, more recently, Business Intelligence (BI), Analytics and many other large-scale systems.
11. Todd Traub, "Wal-Mart Used Technology to Become Supply Chain Leader," *Arkansas Business*, http://www.arkansasbusiness.com/article/85508/wal-mart-used-technology-to-become-supply-chain-leader (accessed July 20, 2013).
12. This is consistent with a similar analysis by Oliner and Sichel (2002), who wrote, "both the use of information technology and the efficiency gains associated with the production of information technology were central factors in [the productivity] resurgence." Oliner, Sichel, and Stiroh (2007) also found that IT was a key factor in this resurgence. Dale Jorgenson, Mun Ho, and Kevin Stiroh, "Will the U.S. Productivity Resurgence Continue?" Federal Reserve Bank of New York: Current Issues in Economics and Finance, December 2004, http://www.newyorkfed.org/research/current_issues/ci10-13/ci10-13.html.

Susan Housman, an economist at the Upjohn Institute has argued that the enormous productivity gains of the computer producing industries unfairly skew the productivity of the manufacturing sector (http://www.minneapolisfed.org/publications_papers/pub_display.cfm?id=4982). She says, "The computer industry is small—it only accounts for about 12 percent of manufacturing's value added. But it has an outsized effect on manufacturing statistics. . . . But we find that without the computer industry, growth in manufacturing real value added falls by two-thirds and productivity growth falls by almost half. It doesn't look like a strong sector without computers." However, we see the glass as half-full, and welcome the contribution of computers even as other sectors lag.

13. See K. J. Stiroh, "Information Technology and the U.S. Productivity Revival: What Do the Industry Data Say?," *American Economic Review* 92, no. 5 (2002): 1559–76; and D. W. Jorgenson, M. S. Ho, and J. D. Samuels, "Information Technology and U.S. Productivity Growth: Evidence from a Prototype Industry Production Account," *Journal of Productivity Analysis*, 36, no. 2 (2011): 159–75, especially table 5, which shows the total factor productivity growth was about ten times higher in IT-using sectors than in sectors that did not use IT extensively.

14. See E. Brynjolfsson and L. M. Hitt, "Computing Productivity: Firm-level Evidence," *Review of Economics and Statistics* 85, no. 4 (2003): 793–808. Similarly, Stanford University's Nicholas Bloom, Harvard University's Rafaela Sadun, and the London School of Economics' John Van Reenen found that American firms were particularly adept at implementing management practices that maximized the value of IT, and this lead to measurable productivity improvements, as documented. See N. Bloom, R. Sadun, and J. Van Reenen "Americans Do IT Better: U.S. Multinationals and the Productivity Miracle (No. w13085)," National Bureau of Economic Research, 2007.

15. Andrew McAfee, "Pharmacy Service Improvement at CVS (A)," *Harvard Business Review*, Case Study, 2005, http://hbr.org/product/pharmacy-service-improvement-at-cvs-a/an/606015-PDF-ENG.

16. Erik Brynjolfsson, Lorin Hitt, and Shinkyu Yang, "Intangible Assets: Computers and Organizational Capital," *Brookings Papers on Economic Activity*, 2002, http://ebusiness.mit.edu/research/papers/138_Erik_Intangible_Assets.pdf.

17. More details can be found in Erik Brynjolfsson and Adam Saunders, *Wired for Innovation: How Information Technology Is Reshaping the Economy* (Cambridge, MA; London: MIT Press, 2013).

18. According to the U.S. Bureau of Labor Statistics, productivity growth averaged 2.4 percent between 2001 and 2010, 2.3 percent between 1991 and 2000, 1.5 percent between 1981 and 1990, and 1.7 percent between 1971 and 1980.

Chapter 8 BEYOND GDP

1. Joel Waldfogel, "Copyright Protection, Technological Change, and the Quality of New Products: Evidence from Recorded Music Since Napster," Working Paper (National Bureau of Economic Research, October 2011), http://www.nber.org/papers/w17503.

2. Albert Gore, *The Future: Six Drivers of Global Change* (New York: Random House, 2013), p. 45.

3. The English Wikipedia has over 2.5 billion words, which is over fifty times as many as *Encyclopaedia Britannica*. "Wikipedia: Size Comparisons," *Wikipedia, the Free Encyclopedia*, July 4, 2013, http://en.wikipedia.org/w/index.php?title=Wikipedia:Size_comparisons&oldid=562880212 (accessed August 17, 2013).

4. Actually, 90 percent of apps on smartphones are now free. Alex Cocotas, "Nine Out Of Ten Apps On Apple's App Store Are Free," *Business Insider*, July 19, 2013, http://www.businessinsider.com/nine-out-of-10-apps-are-free-2013-7#ixzz2cojAAOCy (accessed August 17, 2013).
5. Cannibalization of SMS services by free over-the-top (OTT) service is estimated to cost telephone companies over $30 billion in 2013, according to the analyst group Ovum. Graeme Philipson, "Social Messaging to Cost Telcos $30 Billion in Lost SMS Revenues," *IT Wire*, May 2, 2013, http://www.itwire.com/it-industry-news/strategy/59676-social-messaging-to-cost-telcos-$30-billion-in-lost-sms-revenues (accessed August 17, 2013). In theory, the hardworking statisticians at the Bureau of Economic Analysis try to account for quality-adjusted price changes. In practice, this works for small changes but not for highly disruptive introductions of new products and services. What's more, sometimes increases in GDP reflect declines in our well-being. For instance, an increase in crime might prompt more spending on burglar alarms, police services, and prisons. Every dollar spent on these activities increases GDP, but of course the nation would be better off with less crime and less need for this kind of spending.
6. See http://archive.org/stream/catalogno12400sear#page/370/mode/2up (accessed September 15, 2013).
7. Try the 1912 Sears catalog (p. 873), where it's priced at just 72 cents; see http://archive.org/stream/catalogno12400sear#page/872/mode/2up.
8. It turns out that you get a slightly different answer depending on whether you try to replicate the "happiness" that you had in 1993 using the 2013 catalog, or replicate the happiness of the 2013 catalog using the 1993 catalog. Technically this is the difference between what economists call the Paasche and Laspeyres Price indexes. An alternative is to continually adjust the basket of goods being considered, which is the approach used in so-called chained price indexes. The choice of price index, while subtle, can lead to hundreds of billions of dollars in differences over time, as in the case of indexing Social Security payments for changes in the cost of living.
9. In principle, when the exact same good is available for a lower price, the nominal GDP would fall, but the real GDP would not, with the difference being reflected in the price index. In practice, changes in consumption like this are not picked up in changes in price indexes, and thus official numbers for both nominal and real GDP fall.
10. Erik Brynjolfsson, "The Contribution of Information Technology to Consumer Welfare," *Information Systems Research* 7, no. 3 (1996): 281–300, doi:10.1287/isre.7.3.281.
11. Erik Brynjolfsson and Joo Hee Oh, "The Attention Economy: Measuring the Value of Free Goods on the Internet," in NBER Conference on the Economics of Digitization, Stanford, 2012, http://conference.nber.org/confer//2012/EoDs12/Brynjolfsson_Oh.pdf.
12. Hal Varian, "Economic Value of Google," March 29, 2011, http://cdn.oreillystatic.com/en/assets/1/event/57/The%20Economic%20Impact%20of%20Google%20Presentation.pdf (accessed August 23, 2013). Yan Chen, Grace YoungJoo Jeon, and Yong-Mi Kim, "A Day without a Search Engine: An Experimental Study of Online and Offline Search," http://yanchen.people.si.umich.edu/.

13. Emil Protalinski, "10.5 Billion Minutes Spent on Facebook Daily, Excluding Mobile," *ZDNet*, http://www.zdnet.com/blog/facebook/10-5-billion-minutes-spent-on-facebook-daily-excluding-mobile/11034 (accessed July 23, 2013).

14. Daniel Weld, "Internet Enabled Human Computation," July 22, 2013, Slide 49, https://docs.google.com/viewer?a=v&q=cache:HKa8bKFJkRQJ:www.cs.washington.edu/education/courses/cse454/10au/slides/13-hcomp.ppt+facebook+hours+panama+canal+ahn&hl=en&gl=us&pid=bl&srcid=ADGEESjO16Vz-Mrtg5P2gFvRC82qOoJvsHNVmr56N1XbswDpmqoxb1pUMLoJacAgvNdPRk5OCU0gPCjLbf_3SIvu4oiqCYAqywUkC18VLBdwiE2SwTQrGJXOxuxZFpu_gy6JrmzAtri0&sig=AHIEtbQnKVDd9ybDuAJQJMIMhD8R_oNt8Q.

15. For a good overview, see Clive Thompson, "For Certain Tasks, the Cortex Still Beats the CPU," *Wired*, June 25, 2007.

16. National Science Foundation, "Industry, Technology, and the Global Marketplace," *Science and Engineering Indicators 2012*, 2012, http://www.nsf.gov/statistics/seind12/c6/c6h.htm#s2 (accessed July 27, 2013).

17. Michael Luca, "Reviews, Reputation, and Revenue: The Case of Yelp.com," Harvard Business School Working Paper (Harvard Business School, 2011), http://ideas.repec.org/p/hbs/wpaper/12-016.html (accessed September 12, 2013).

18. Ralph Turvey, "Review of: Toward a More Accurate Measure of the Cost of Living: Final Report to the Senate Finance Committee from the Advisory Committee to Study the Consumer Price Index. by Michael J. Boskin; Ellen R. Dullberger; Robert J. Gordon," *Economic Journal* 107, no. 445 (1997): 1913–15, doi:10.2307/2957930.

19. Jonathan Rothwell et al., "Patenting Prosperity: Invention and Economic Performance in the United States and Its Metropolitan Areas," February 2013, http://www.brookings.edu/research/reports/2013/02/patenting-prosperity-rothwell (accessed September 12, 2013).

20. See Carol Corrado, Chuck Hulten, and Dan Sichel, "Intangible Capital and Economic Growth," NBER Working Paper No. 11948, 2006, http://www.nber.org/papers/w11948.

21. Erik Brynjolfsson, Lorin Hitt, and Shinkyu Yang, "Intangible Assets: Computers and Organizational Capital," Brookings Papers on Economic Activity, 2002, http://ebusiness.mit.edu/research/papers/138_Erik_Intangible_Assets.pdf (accessed August 18, 2013); Erik Brynjolfsson and Lorin M. Hitt, "Computing Productivity: Firm-Level Evidence," SSRN Scholarly Paper (Rochester, NY: Social Science Research Network, 2003), http://papers.ssrn.com/abstract=290325.

22. Rick Burgess, "One Minute on the Internet: 640TB Data Transferred, 100k Tweets, 204 Million E-mails Sent," *TechSpot*, http://www.techspot.com/news/52011-one-minute-on-the-internet-640tb-data-transferred-100k-tweets-204-million-e-mails-sent.html (accessed July 23, 2013).

23. "Facebook Newsroom," http://newsroom.fb.com/content/default.aspx?NewsAreaId=21 (accessed July 23, 2013).

24. Dale Jorgenson and Barbara Fraumeni, "The Accumulation of Human and Nonhuman Capital, 1948–84," in *The Measurement of Saving, Investment, and Wealth* (Chi-

cago, IL: University of Chicago Press for National Bureau of Economic Research, 1989), p. 230, http://www.nber.org/chapters/c8121.pdf.
25. Adam Smith, *An Inquiry into the Nature and Causes of the Wealth of Nations*, ed. Edwin Cannan (Library of Economics and Liberty, 1904), http://www.econlib.org/library/Smith/smWN20.html (accessed September 23, 2013).
26. Ana Aizcorbe, Moylan Carol, and Robbins Carol, "Toward Better Measurement of Innovation and Intangibles," BEA Briefing, January 2009, http://www.bea.gov/scb/pdf/2009/01%20January/0109_innovation.pdf.
27. As quoted in "GDP: One of the Great Inventions of the 20th Century," January 2000 Survey of Current Business, http://www.bea.gov/scb/account_articles/general/0100od/maintext.htm.
28. Joseph E. Stiglitz, "GDP Fetishism," *Project Syndicate*, http://www.project-syndicate.org/commentary/gdp-fetishism (accessed July 23, 2013).
29. "Human Development Index (HDI)," *Human Development Reports*, 2012, http://hdr.undp.org/en/statistics/hdi/ (accessed July 23, 2013).
30. "Policy—A Multidimensional Approach," *Oxford Poverty & Human Development Initiative*, 2013, http://www.ophi.org.uk/policy/multidimensional-poverty-index/.
31. "DHS Overview," *Measure DHS: Demographic and Health Surveys*, 2013, http://www.measuredhs.com/What-We-Do/Survey-Types/DHS.cfm (accessed September 11, 2013).
32. Joseph Stiglitz, Amartya Sen, and Jean-Paul Fitoussi, "Report by the Commission on the Measurement of Economic Performance and Social Progress," *Council on Foreign Relations*, August 25, 2010, http://www.cfr.org/world/report-commission-measurement-economic-performance-social-progress/p22847 (accessed August 9, 2013).
33. See the Social Progress Index at http://www.socialprogressimperative.org/data/spi.
34. See the Well-Being Index at http://www.well-beingindex.com/.
35. See the MIT Billion Prices Project at http://bpp.mit.edu.
36. See, for example, Hyunyoung Choi and Hal Varian, "Predicting the Present with Google Trends," *Google Inc.*, April 10, 2009, http://static.googleusercontent.com/external_content/untrusted_dlcp/www.google.com/en/us/googleblogs/pdfs/google_predicting_the_present.pdf (accessed September 11, 2013); Lynn Wu and Erik Brynjolfsson, "The Future of Prediction: How Google Searches Foreshadow Housing Prices and Sales," SSRN Scholarly Paper (Rochester, NY: Social Science Research Network, 2013), http://papers.ssrn.com/abstract=2022293.

Chapter 9 THE SPREAD

1. Jonathan Good, "How Many Photos Have Ever Been Taken?," *1000memories*, September 15, 2011, http://blog.1000memories.com/94-number-of-photos-ever-taken-digital-and-analog-in-shoebox (accessed August 10, 2013).
2. Ibid.

3. Tomi Ahonen, "Celebrating 30 Years of Mobile Phones, Thank You NTT of Japan," *Communities Dominate Brands*, November 13, 2009, http://communities-dominate.blogs.com/brands/2009/11/celebrating-30-years-of-mobile-phones-thank-you-ntt-of-japan.html (accessed September 11, 2013).
4. Good, "How Many Photos Have Ever Been Taken?"
5. Craig Smith, "By the Numbers: 12 Interesting Instagram Stats," *Digital Marketing Ramblings . . .* , June 23, 2013, http://expandedramblings.com/index.php/important-instagram-stats/ (accessed August 10, 2013).
6. Leena Rao, "Facebook Will Grow Headcount Quickly In 2013 To Develop Money-Making Products, Total Expenses Will Jump By 50 Percent," *TechCrunch*, January 30, 2013, http://techcrunch.com/2013/01/30/zuck-facebook-will-grow-headcount-quickly-in-2013-to-develop-future-money-making-products/ (accessed August 10, 2013).
7. Brad Stone and Ashlee Vance, "Facebook's 'Next Billion': A Q&A With Mark Zuckerberg," *Bloomberg Businessweek*, October 4, 2012, http://www.businessweek.com/articles/2012-10-04/facebooks-next-billion-a-q-and-a-with-mark-zuckerberg (accessed September 11, 2013).
8. "Kodak's Growth and Decline: A Timeline," *Rochester Business Journal*, January 19, 2012, http://www.rbj.net/print_article.asp?aID=190078.
9. According to an analysis of 2006 tax returns in the United States by Emmanuel Saez of University of California, Berkeley.
10. In contrast, life expectancy for men and women with more than a high school education increased during this period.
11. Sylvia Allegretto, "The State of Working America's Wealth, 2011," Briefing Paper No. 292, Economic Policy Institute, Washington, D.C.
12. See for example, Josh Bivens, "Inequality, Exhibit A: Walmart and the Wealth of American Families," *Working Economics*, Economic Policy Institute blog, http://www.epi.org/blog/inequality-exhibit-wal-mart-wealth-american/ (accessed September 17, 2013).
13. Luisa Kroll, "Inside the 2013 Forbes 400: Facts and Figures On America's Richest," *Forbes*, September 16, 2013, http://www.forbes.com/sites/luisakroll/2013/09/16/inside-the-2013-forbes-400-facts-and-figures-on-americas-richest/ (accessed September 16, 2013).
14. About one-third of the difference reflected technical differences in the way output prices are calculated when used in productivity calculations versus the consumer prices used in calculating income. In addition, about 12 percent was due to the growth of nonwage benefits such as health care. See Lawrence Mishel, "The Wedges between Productivity and Median Compensation Growth," Economic Policy Institute, April 26, 2012, http://www.epi.org/publication/ib330-productivity-vs-compensation/. When looking at household income, about 20 percent of the decline reflects the fact that households are somewhat smaller than they were thirty years ago.
15. Data from the Organization for Economic Cooperation and Development (OECD)

show that income inequality increased in seventeen of twenty-two nations including Mexico, the United States, Israel, United Kingdom, Italy, Australia, New Zealand, Japan, Canada, Germany, Netherlands, Luxembourg, Finland, Sweden, Czech Republic, Norway, and Denmark. See "An Overview of Growing Income Inequalities in the OECD Countries: Main Findings," from the OECD, 2011, http://www.oecd.org/social/soc/49499779.pdf.

16. See, for instance, Robert M. Solow, "Technical Change and the Aggregate Production Function," *Review of Economics and Statistics* 39, no. 3 (1957): 312–20, doi:10.2307/1926047.

17. See David H. Autor, Lawrence F. Katz, and Alan B. Krueger, "Computing Inequality: Have Computers Changed the Labor Market?," Working Paper (National Bureau of Economic Research, March 1997), http://www.nber.org/papers/w5956; F. Levy and R. J. Murnane, *The New Division of Labor: How Computers Are Creating the Next Job Market* (Princeton, NJ: Princeton University Press, 2012); D. Autor, "The Polarization of Job Opportunities in the U.S. Labor Market," The Brookings Institution, http://www.brookings.edu/research/papers/2010/04/jobs-autor (accessed August 10, 2013); and Daron Acemoglu and David Autor, "Skills, Tasks and Technologies: Implications for Employment and Earnings," Working Paper (National Bureau of Economic Research, June 2010), http://www.nber.org/papers/w16082.

18. Daron Acemoglu and David Autor, "Skills, Tasks and Technologies: Implications for Employment and Earnings," *Handbook of Labor Economics* 4 (2011): 1043–1171.

19. See "Digest of Education Statistics, 1999," National Center for Education Statistics, http://nces.ed.gov/programs/digest/d99/d99t187.asp (accessed August 10, 2013).

20. See T. F. Bresnahan, E. Brynjolfsson, and L. M. Hitt, "Information Technology, Workplace Organization, and the Demand for Skilled Labor: Firm-level Evidence," *Quarterly Journal of Economics*, 117, no. 1 (2002): 339–76. See also E. Brynjolfsson, L. M. Hitt, and S. Yang, "Intangible Assets: Computers and Organizational Capital," Brookings Papers on Economic Activity, 2002, pp. 137–98.

21. See Brynjolfsson, Hitt, and Yang, "Intangible Assets: Computers and Organizational Capital," and Erik Brynjolfsson, David Fitoussi, and Lorin Hitt, "The IT Iceberg: Measuring the Tangible and Intangible Computing Assets," Working Paper (October 2004).

22. E. Brynjolfsson and L. M. Hitt, "Computing Productivity: Firm-level Evidence," *Review of Economics and Statistics* 8, no. 4 (2003): 793–808.

23. Timothy F. Bresnahan, Erik Brynjolfsson, and Lorin M. Hitt, "Information Technology, Workplace Organization, and the Demand for Skilled Labor: Firm-Level Evidence," *Quarterly Journal of Economics* 117, no. 1 (2002): 339–76, doi:10.1162/003355302753399526.

24. Reengineering consultants like to tell the story of how, in the seventeenth century, cows roamed around Boston Common and the neighboring areas. Over time, these cow paths became well-worn, and as shops and homes were constructed, people used the same paths for their carts and carriages. Eventually cobblestones were installed, and by the twentieth century most of the paths had been paved over

with asphalt, with no more cows to be seen. As anyone who's tried to drive in Boston can appreciate, having traffic flow designed by cows may not be the best way to lay out a modern city.

25. See David Autor, "The Polarization of Job Opportunities in the U.S. Labor Market," Brookings Institution (April 2010), http://www.brookings.edu/research/papers/2010/04/jobs-autor (accessed August 10, 2013); and Daron Acemoglu and David Autor, "Skills, Tasks and Technologies: Implications for Employment and Earnings," Working Paper (National Bureau of Economic Research, June 2010), http://www.nber.org/papers/w16082.

26. See N. Jaimovich and H. E. Siu, "The Trend is the Cycle: Job Polarization and Jobless Recoveries (No. w18334)," National Bureau of Economic Research, 2012.

27. As Hans Moravec put it, "it is comparatively easy to make computers exhibit adult level performance on intelligence tests or playing checkers, and difficult or impossible to give them the skills of a one-year-old when it comes to perception and mobility." Hans Moravec, *Mind Children: The Future of Robot and Human Intelligence* (Cambridge, MA: Harvard University Press, 1988).

28. See chapter 6 in Jonathan Schaeffer, *One Jump Ahead: Computer Perfection at Checkers* (New York: Springer, 2009), http://public.eblib.com/EBLPublic/PublicView.do?ptiID=418209.

29. Quoted in Daniel Crevier, *AI: The Tumultuous History of the Search for Artificial Intelligence* (New York: Basic Books, 1993), p. 108.

30. Jack Copeland, "A Brief History of Computing," June 2000, http://www.alanturing.net/turing_archive/pages/Reference%20Articles/BriefHistofComp.html.

31. The mobile phone chess game Pocket Fritz won the Copa Mercosur tournament in Buenos Aires, Argentina, in 2009. "Breakthrough Performance by Pocket Fritz 4 in Buenos Aires," *Chess News*, http://en.chessbase.com/Home/TabId/211/PostId/4005719/breakthrough-performance-by-pocket-fritz-4-in-buenos-aires.aspx (accessed August 10, 2013).

32. Steve Musil, "Foxconn Reportedly Installing Robots to Replace Workers" *CNET*, November 13, 2012, http://news.cnet.com/8301-1001_3-57549450-92/foxconn-reportedly-installing-robots-to-replace-workers/ (accessed November 13, 2012).

33. Rod Brooks gave four dollars per hour as the approximate cost of Baxter in response to a question at the Techonomy 2012 Conference in Tucson, Arizona, on November 12, 2012, during a panel discussion with Andrew McAfee.

34. Karl Marx, *Capital: A Critique of Political Economy* (New York: Modern Library, 1906), pp. 708–9.

35. See Dale Jorgenson, *A New Architecture for the U.S. National Accounts* (Chicago, IL: University of Chicago Press, 2006).

36. Susan Fleck, John Glaser, and Shawn Sprague, "The Compensation-Productivity Gap: A Visual Essay," *Monthly Labor Review* (January 2011), http://www.bls.gov/opub/mlr/2011/01/art3full.pdf, p. 57-69.

37. L. Karabarbounis and B. Neiman, "The Global Decline of the Labor Share (No. w19136)," National Bureau of Economic Research, 2013.

38. See http://w3.epi-data.org/temp2011/BriefingPaper324_FINAL %283%29.pdf.
39. See http://blogs.wsj.com/economics/2011/09/28/its-man-vs-machine-and-man-is-losing/.
40. See, e.g., Lucian A. Bebchuk and Yaniv Grinstein, "The Growth of Executive Pay," *Oxford Review of Economic Policy* 21 (2005): 283–303; Harvard Law and Economics Discussion Paper No. 510. Available at SSRN, http://papers.ssrn.com/abstract=648682 (accessed August 10, 2013).

Chapter 10 THE BIGGEST WINNERS: STARS AND SUPERSTARS

1. *Nike—You Don't Win Silver, You Lose Gold*, 2012, http://www.youtube.com/watch?v=ZnLCeXMHzBs&feature=youtube_gdata_player.
2. In most cases, the winner does not literally take all of the market. Perhaps 'winner-take-most' would be a more accurate description. But for better or worse, in the competition for concept-naming among economists, 'winner-take-all' has won almost all the market share, so that's what we will use.
3. Emmanuel Saez, "Striking It Richer: The Evolution of Top Incomes in the United States," January 23, 2013, http://elsa.berkeley.edu/~saez/saez-UStopincomes-2011.pdf.
4. "Why The Haves Have So Much : NPR," *NPR.org*, October 29, 2011, http://www.npr.org/2011/10/29/141816778/why-the-haves-have-so-much (accessed August 11, 2013).
5. Alex Tabarrok, "Winner Take-All Economics," *Marginal Revolution*, September 13, 2010, http://marginalrevolution.com/marginalrevolution/2010/09/winner-take-all-economics.html.
6. Steven N. Kaplan and Joshua Rauh, "It's the Market: The Broad-Based Rise in the Return to Top Talent," *Journal of Economic Perspectives* 27, no. 3 (2013): 35–56.
7. David Streitfeld, "As Boom Lures App Creators, Tough Part Is Making a Living," *New York Times*, November 17, 2012, http://www.nytimes.com/2012/11/18/business/as-boom-lures-app-creators-tough-part-is-making-a-living.html.
8. Heekyung Kim and Erik Brynjolfsson, "CEO Compensation and Information Technology," *ICIS 2009 Proceedings*, January 1, 2009, http://aisel.aisnet.org/icis2009/38.
9. See Xavier Gabaix and Augustin Landier, "Why Has CEO Pay Increased so Much?," SSRN Scholarly Paper (Rochester, NY: Social Science Research Network, May 8, 2006), http://papers.ssrn.com/abstract=901826.
10. Robert H. Frank and Philip J. Cook, *The Winner-take-all Society: Why the Few at the Top Get so Much More Than the Rest of Us* (New York: Penguin Books, 1996).
11. Sherwin Rosen, "The Economics of Superstars," *American Economic Review* 71, no. 5 (1981): 845–58, doi:10.2307/1803469.
12. D. Rush, "Google buys Waze map app for $1.3bn," *Guardian* (UK), June 11, 2013, http://www.theguardian.com/technology/2013/jun/11/google-buys-waze-maps-billion.
13. You can see the video and data on number of viewings at https://www.youtube.com/watch?v=OYpwAtnywTk.

14. See Roy Jones and Haim Mendelson on this point: "Information Goods vs. Industrial Goods: Cost Structure and Competition," *Management Science* 57, no. 1 (2011): 164–76, doi:10.1287/mnsc.1100.1262.
15. Very low marginal costs can also make massive bundling more profitable. That's one reason why cable TV tends to be sold in bundles rather than à la carte, and why Microsoft Office was able to win market share from more focused products. Bundling benefits both superstars and niche providers by creating a more complete product offering and increasing sales to consumers with different opinions about the values of the bundled products. But markets in which bundling is common also tend to be winner-take-all markets. See Yannis Bakos and Erik Brynjolfsson, *Management Science* 45, no. 12 (1999); Yannis Bakos and Erik Brynjolfsson, "Bundling and Competition on the Internet," *Marketing Science* 19, no. 1 (2000): 63–82, doi:10.1287/mksc.19.1.63.15182.
16. See Michael D. Smith and Erik Brynjolfsson, "Consumer Decision-making at an Internet Shopbot: Brand Still Matters," *NBER* (December 1, 2001): 541–58.
17. Catherine Rampell, "College Degree Required by Increasing Number of Companies," *New York Times*, February 19, 2013, http://www.nytimes.com/2013/02/20/business/college-degree-required-by-increasing-number-of-companies.html.
18. We discuss this more in our article "Investing in the IT That Makes a Competitive Difference," July 2008, http://hbr.org/2008/07/investing-in-the-it-that-makes-a-competitive-difference.
19. Alfred Marshall, *Principles of Economics*, 8th edition, New York: Macmillan, 1947, p. 685.
20. See, e.g., http://www.koomey.com/books.html or http://www.johntreed.com/FCM.html.
21. We discuss this in more detail in a *Harvard Business Review* article (A. McAfee and E. Brynjolfsson, "Investing in the IT That Makes a Competitive Difference: Studies of Corporate Performance Reveal a Growing Link between Certain Kinds of Technology Investments and Intensifying Competitiveness," *Harvard Business Review* [2006]: 98–103) and a research paper (E. Brynjolfsson, A. McAfee, M. Sorell, and F. Zhu, "Scale without Mass: Business Process Replication and Industry Dynamics," MIT Center for Digital Business Working Paper, 2008).
22. More formally, power laws are characterized by the formula $f(x) = ax^k$. For instance, the sales of a book at Amazon, $f(x)$ are a function of the rank of the book, x, raised to the power k. A nice characteristic of power laws is that they form a straight line when graphed on a log-log scale, with the slope of the line given by the exponent, k.
23. Erik Brynjolfsson, Yu Jeffrey Hu, and Michael D. Smith, "Consumer Surplus in the Digital Economy: Estimating the Value of Increased Product Variety at Online Booksellers," SSRN Scholarly Paper (Rochester, NY: Social Science Research Network, June 1, 2003), http://papers.ssrn.com/abstract=400940.
24. In other words, so-called "black swan" events are more common if the underlying distribution is a power law than if it's a normal distribution.

25. Technically, the bulk of the incomes are best described by a log-normal distribution, a variant of the traditional normal distribution, while the best fit for the top incomes is a power law.
26. Presentation by Kim Taipale at the 21st Annual Aspen Institute Roundtable on Information Technology, August 1, 2013.
27. If you're a nerd, you may know that in some cases the mean of the power-law distribution is actually infinite. Specifically, when the exponent in the distribution (*k* in the above equation) is less than two, the mean of the distribution is infinite.
28. See "Dollars and Sense Part Two: MLB Player Salary Analysis," *Purple Row*, http://www.purplerow.com/2009/4/23/848870/dollars-and-sense-part-two-mlb (accessed August 10, 2013). The disparity would likely be even greater if one considered the endorsement deals that the superstars get.

Chapter 11 IMPLICATIONS OF THE BOUNTY AND THE SPREAD

1. "The World's Billionaires: 25th Anniversary Timeline," *Forbes*, 2012, http://www.forbes.com/special-report/2012/billionaires-25th-anniversary-timeline.html (accessed August 7, 2013); "Income, Poverty and Health Insurance Coverage in the United States: 2011," U.S. Census Bureau Public Information Office, September 12, 2012, http://www.census.gov/newsroom/releases/archives/income_wealth/cb12-172.html (accessed August 9, 2013).
2. N. G. Mankiw, "Defending the One Percent," *Journal of Economic Perspectives*, June 8, 2013, http://scholar.harvard.edu/files/mankiw/files/defending_the_one_percent_0.pdf.
3. Felix Salmon, "Krugman vs. Summers: The Debate," *Reuters Blogs—Felix Salmon*, November 15, 2011, http://blogs.reuters.com/felix-salmon/2011/11/15/krugman-vs-summers-the-debate/ (accessed August 10, 2013).
4. Donald J. Boudreaux and Mark J. Perry, "The Myth of a Stagnant Middle Class," *Wall Street Journal*, January 23, 2013, http://online.wsj.com/article/SB10001424127887323468604578249723138161566.html.
5. Mark J. Perry, "Thanks to Technology, Americans Spend Dramatically Less on Food Than They Did 3 Decades Ago," *AEIdeas*, April 7, 2013, http://www.aei-ideas.org/2013/04/technology-innovation-and-automation-have-lowered-the-cost-of-our-food-and-improved-the-lives-of-all-americans/.
6. Scott Winship, "Myths of Inequality and Stagnation," The Brookings Institution, March 27, 2013, http://www.brookings.edu/research/opinions/2013/03/27-inequality-myths-winship (accessed August 10, 2013).
7. Jared Bernstein, "Three Questions About Consumer Spending and the Middle Class," Bureau of Labor Statistics, June 22, 2010, http://www.bls.gov/cex/duf2010bernstein1.pdf.
8. Annamaria Lusardi, Daniel J. Schneider, and Peter Tufano, "Financially Fragile Households: Evidence and Implications," Working Paper (National Bureau of Economic Research, May 2011), http://www.nber.org/papers/w17072.

9. Jason Matthew DeBacker et al., "Rising Inequality: Transitory or Permanent? New Evidence from a Panel of U.S. Tax Returns 1987-2006," SSRN Scholarly Paper (Rochester, NY: Social Science Research Network, January 2, 2012), http://papers.ssrn.com/abstract=1747849.

10. Robert D. Putnam, "Crumbling American Dreams," *Opinionator, New York Times* blog, August 3, 2013, http://opinionator.blogs.nytimes.com/2013/08/03/crumbling-american-dreams/.

11. "Repairing the Rungs on the Ladder," *The Economist*, February 9, 2013, http://www.economist.com/news/leaders/21571417-how-prevent-virtuous-meritocracy-entrenching-itself-top-repairing-rungs (accessed August 10, 2013).

12. Daron Acemoglu and James A. Robinson, "The Problem with U.S. Inequality," *Huffington Post*, March 11, 2012, http://www.huffingtonpost.com/daron-acemoglu/us-inequality_b_1338118.html (accessed August 13, 2013).

13. John Bates Clark, *Essentials of Economic Theory as Applied to Modern Problem of Industry and Public Policy*, (London: Macmillan, 1907), p. 45.

14. W. M. Leiserson, *The Problem of Unemployment Today* 31, *Political Science Quarterly* (1916), http://archive.org/details/jstor-2141701, p. 12.

15. John Maynard Keynes, *Essays in Persuasion* (New York: W. W. Norton & Company, 1963), p. 358.

16. Linus Pauling, *The Triple Revolution* (Santa Barbara, CA: Ad Hoc Committee on the Triple Revolution, 1964), http://osulibrary.oregonstate.edu/specialcollections/coll/pauling/peace/papers/1964p.7-02.html.

17. Wassily Leontief, "National Perspective: The Definition of Problems and Opportunities," *The Long-Term Impact of Technology on Employment and Unemployment* (National Academy of Engeneering, 1983): 3–7.

18. Richard M. Cyert and David C. Mowery, eds., *Technology and Employment: Innovation and Growth in the U.S. Economy* (National Academies Press, 1987), http://www.nap.edu/catalog.php?record_id=1004.

19. Raghuram Rajan, Paolo Volpin, and Luigi Zingales, "The Eclipse of the U.S. Tire Industry," Working Paper (Center for Economic Studies, U.S. Census Bureau, 1997), http://ideas.repec.org/p/cen/wpaper/97-13.html.

20. William D. Nordhaus, "Do Real Output and Real Wage Measures Capture Reality? The History of Lighting Suggests Not," Cowles Foundation Discussion Paper (Cowles Foundation for Research in Economics, Yale University, 1994), http://ideas.repec.org/p/cwl/cwldpp/1078.html.

21. In one paper, Erik estimated that the elasticity of demand for computer hardware was about 1.1, implying that each 1 percent increase in price led to a 1.1 percent increase in demand, so as a result total spending increased as technology made computers more efficient. See Erik Brynjolfsson, "The Contribution of Information Technology to Consumer Welfare," *Information Systems Research* 7, no. 3 (1996): 281–300.

22. This is an example of Say's Law, which states that demand and supply are always kept in balance.

23. John Maynard Keynes, "Economic Possibilities for Our Grandchildren," *Keynes on Possibilities*, 1930, http://www.econ.yale.edu/smith/econ116a/keynes1.pdf.
24. Tim Kreider, "The 'Busy' Trap," *Opinionator*, June 30, 2012, http://opinionator.blogs.nytimes.com/2012/06/30/the-busy-trap/.
25. Nobel Prize winner Joe Stiglitz has argued that rapid automation of agriculture, such as via gasoline-engine tractors, is part of the explanation for the high unemployment of the 1930s. See Joseph E. Stiglitz, *The Price of Inequality: How Today's Divided Society Endangers Our Future* (New York: W. W. Norton & Company, 2013).
26. Wassily Leontief, "Technological Advance, Economic Growth, and the Distribution of Income," *Population and Development Review* 9, no. 3 (September 1, 1983), 403–10.
27. Michael Spence, *The Next Convergence: The Future of Economic Growth in a Multispeed World* (New York: Macmillan, 2011).
28. D. Autor, D. Dorn, and G. H. Hanson, "The China Syndrome: Local Labor Market Effects of Import Competition in the United States," *American Economic Review* (forthcoming, December 2013).
29. J. Banister and G. Cook, "China's Employment and Compensation Costs in Manufacturing through 2008," *Monthly Labor Review* 134, no. 3 (2011): 39–52. A closer look at the Chinese statistics suggest that the classification methods have changed somewhat over time, so the exact changes in employment may be somewhat different than reported, but the general trend seems clear.

Chapter 12 LEARNING TO RACE *WITH* MACHINES

1. "Computers Are Useless. They Can Only Give You Answers," *Quote Investigator*, November 5, 2011, http://quoteinvestigator.com/2011/11/05/computers-useless/.
2. D. T. Max, "The Prince's Gambit," *The New Yorker*, March 21, 2011, http://www.newyorker.com/reporting/2011/03/21/110321fa_fact_max.
3. Garry Kasparov, "The Chess Master and the Computer," *New York Review of Books*, February 11, 2010, http://www.nybooks.com/articles/archives/2010/feb/11/the-chess-master-and-the-computer/.
4. "Chess Quotes," http://www.chessquotes.com/player-karpov (accessed September 12, 2013).
5. Kasparov, "The Chess Master and the Computer."
6. Evan Esar, *20,000 Quips & Quotes* (Barnes and Noble, 1995), p. 654.
7. Kevin Kelly, "Better than Human: Why Robots Will—and Must—Take Our Jobs," *Wired*, December 24, 2012.
8. Zara's approach is described in more detail in a Harvard Business Case Study by Andy and two colleagues: Andrew McAfee, Vincent Dessain, and Anders Sjöman, "Zara: IT for Fast Fashion," Harvard Business School, 2007 (Case number 604081-PDF-ENG).
9. John Timbs, "The Mirror of Literature, Amusement, and Instruction (London: John Limbird, 1825)," p. 75.

10. Sugata Mitra, "Build a School in the Cloud," *TED*, video on TED.com, February 2013, http://www.ted.com/talks/sugata_mitra_build_a_school_in_the_cloud.html.
11. Ibid.
12. Peter Sims, "The Montessori Mafia," *Wall Street Journal*, April 5, 2011, http://blogs.wsj.com/ideas-market/2011/04/05/the-montessori-mafia/.
13. Richard Arum and Josipa Roksa, *Academically Adrift: Limited Learning on College Campuses* (Chicago, IL: University of Chicago Press, 2010); Richard Arum, Josipa Roksa, and Esther Cho, "Improving Undergraduate Learning: Findings and Policy Recommendations from the SSRC-CLA Longitudinal Project," Social Science Research Council, 2008, http://www.ssrc.org/publications/view/D06178BE-3823-E011-ADEF-001CC477EC84/.
14. Ernest T. Pascarella and Patrick T. Terenzini, *How College Affects Students: A Third Decade of Research*, 1st ed. (San Francisco: Jossey-Bass, 2005), 602.
15. Michael Noer, "One Man, One Computer, 10 Million Students: How Khan Academy Is Reinventing Education," *Forbes*, November 19, 2012, http://www.forbes.com/sites/michaelnoer/2012/11/02/one-man-one-computer-10-million-students-how-khan-academy-is-reinventing-education/.
16. William J. Bennet, "Is Sebastian Thrun's Udacity the Future of Higher Education?" *CNN*, July 5, 2012, http://www.cnn.com/2012/07/05/opinion/bennett-udacity-education/index.html.
17. David Autor, "The Polarization of Job Opportunities in the U.S. Labor Market: Implications for Employment and Earnings," Brookings Institution, April 2010, http://www.brookings.edu/research/papers/2010/04/jobs-autor.
18. Catherine Rampell, "Life Is O.K., If You Went to College," *Economix* blog, *New York Times*, May 3, 2013, http://economix.blogs.nytimes.com/2013/05/03/life-is-o-k-if-you-went-to-college/.
19. Catherine Rampell, "College Degree Required by Increasing Number of Companies," *New York Times*, February 19, 2013, http://www.nytimes.com/2013/02/20/business/college-degree-required-by-increasing-number-of-companies.html.
20. Meta Brown et al., "Grading Student Loans," *Liberty Street Economics* blog, Federal Reserve Bank of New York, March 5, 2012, http://libertystreeteconomics.newyorkfed.org/2012/03/grading-student-loans.html?utm_source=feedburner&utm_medium=feed&utm_campaign=Feed:+LibertyStreetEconomics+(Liberty+Street+Economics).
21. Tim Hornyak, "Towel-folding Robot Won't Do the Dishes," *CNET*, March 31, 2010, http://news.cnet.com/8301-17938_105-10471898-1.html.
22. Nate Silver, *The Signal and the Noise: Why So Many Predictions Fail—But Some Don't*, 1st ed. (New York: Penguin, 2012).

Chapter 13 POLICY RECOMMENDATIONS

1. "Employment Level," *Economic Research—Federal Reserve Bank of St. Louis* (U.S. Department of Labor, Bureau of Labor Statistics, August 2, 2013), http://research.stlouisfed.org/fred2/series/LNU02000000.

2. Claudia Goldin and Lawrence F. Katz, *The Race Between Education and Technology* (Cambridge, MA: Belknap Press of Harvard University Press, 2010).
3. "PISA 2009 Key Findings," *OECD*, http://www.oecd.org/pisa/pisaproducts/pisa2009/pisa2009keyfindings.htm (accessed August 12, 2013).
4. Martin West, "Global Lessons for Improving U.S. Education," September 29, 2011, http://www.issues.org/28.3/west.html.
5. Marcella Bombardieri, "Professors Take Lessons from Online Teaching," *Boston Globe*, June 9, 2013, http://www.bostonglobe.com/metro/2013/06/08/professors-take-lessons-from-online-teaching/K5XTNA8N1cVGLQ8JJW5PCL/story.html (accessed August 19, 2013).
6. Raj Chetty, John N. Friedman, and Jonah E. Rockoff, "The Long-Term Impacts of Teachers: Teacher Value-Added and Student Outcomes in Adulthood," NBER Working Paper (National Bureau of Economic Research, 2011), http://ideas.repec.org/p/nbr/nberwo/17699.html.
7. Ray Fisman, "Do Charter Schools Work?," *Slate*, May 22, 2013, http://www.slate.com/articles/news_and_politics/the_dismal_science/2013/05/do_charter_schools_work_a_new_study_of_boston_schools_says_yes.single.html (accessed August 12, 2013).
8. Olga Khazan, "Here's Why Other Countries Beat the U.S. in Reading and Math," *Washington Post*, December 11, 2012, http://www.washingtonpost.com/blogs/worldviews/wp/2012/12/11/heres-why-other-countries-beat-the-u-s-in-reading-and-math/ (accessed August 12, 2013).
9. See, for instance, Miles Kimball's praise of the "Knowledge is Power Program": "Confessions of a Supply-Side Liberal," July 23, 2012, http://blog.supplysideliberal.com/post/27813547755/magic-ingredient-1-more-k-12-school (accessed August 12, 2013).
10. B. Holmstrom and P. Milgrom, "Multitask Principal-Agent Analyses: Incentive Contracts, Asset Ownership, and Job Design," *Journal of Law, Economics & Organization* 7, no. 24 (1991).
11. Joseph Alois Schumpeter, *The Theory of Economic Development: An Inquiry Into Profits, Capital, Credit, Interest, and the Business Cycle* (Piscataway, NJ: Transaction Publishers, 1934).
12. Ibid., p. 66.
13. Press Release, "U.S. Job Growth Driven Entirely by Startups, According to Kauffman Foundation Study," Reuters, July 7, 2010, http://www.reuters.com/article/2010/07/07/idUS165927+07-Jul-2010+MW20100707.
14. John Haltiwanger et al., "Business Dynamics Statistics Briefing: Job Creation, Worker Churning, and Wages at Young Businesses," SSRN Scholarly Paper (Rochester, NY: Social Science Research Network, November 1, 2012), http://papers.ssrn.com/abstract=2184328.
15. "Kauffman Index of Entrepreneurial Activity," Ewing Marion Kauffman Foundation, 2012, http://www.kauffman.org/research-and-policy/kauffman-index-of-entrepreneurial-activity.aspx.

16. Vivek Wadhwa, AnnaLee Saxenian, and Francis Daniel Siciliano, "Then and Now: America's New Immigrant Entrepreneurs," Part 7, Stanford Public Law Working Paper No. 2159875; Rock Center for Corporate Governance at Stanford University Working Paper No. 127, SSRN Scholarly Paper (Rochester, NY: Social Science Research Network, October 1, 2012), http://papers.ssrn.com/abstract=2159875.
17. Leora Klapper, Luc Laeven, and Raghuram Rajan, "Entry Regulation as a Barrier to Entrepreneurship," *Journal of Financial Economics* 82, no. 3 (2006): 591–629, doi:10.1016/j.jfineco.2005.09.006.
18. "Research and Development: Essential Foundation for U.S. Competitiveness in a Global Economy," in *A Companion to Science and Engineering Indicators 2008* (National Science Board, January 2008), http://www.nsf.gov/statistics/nsb0803/start.htm.
19. In her new book, *The Entrepreneurial State*, Mariana Mazzucato nicely illustrates this point, noting that each of the core technologies in Apple's breakthrough iPhone were based on government-funded research, including cellular telephony, the Internet, GPS, microchips, capacitive sensors, the touchscreen, and even Siri. See Mariana Mazzucato, *The Entrepreneurial State: Debunking Public vs. Private Sector Myths* (New York: Anthem Press, 2013).
20. If that fact made you worry that you may owe back royalties from your public performance at that restaurant last week, you may be in luck. The two million dollars a year in licensing fees collected by the owner of the "Happy Birthday" copyright is being challenged and may be overturned. See Jacob Goldstein, "This One Page Could End The Copyright War Over 'Happy Birthday,' " NPR, June 17, 2013, http://www.npr.org/blogs/money/2013/06/17/192676099/this-one-page-could-end-the-copyright-war-over-happy-birthday.
21. This list is drawn from Tom Kalil's Grand Challenges presentation. A copy is available at http://www2.itif.org/2012-grand-challenges-kalil.pdf (accessed August 9, 2013). See also "Implementation of Federal Prize Authority: Progress Report" by the U.S. Office of Science and Technology Policy, March 2012, available at http://www.whitehouse.gov/sites/default/files/microsites/ostp/competes_report_on_prizes_final.pdf (accessed September 18, 2013).
22. For a detailed list, see the appendix of McKinsey and Company, "And the Winner Is . . . " Research Report, 2009, http://mckinseyonsociety.com/downloads/reports/Social-Innovation/And_the_winner_is.pdf (accessed September 18, 2013).
23. "2013 Report Card for America's Infrastructure," ASCE, 2013, http://www.infrastructurereportcard.org/a/#p/home (accessed August 12, 2013).
24. See Matthew Yglesias, "The Collapse of Public Investment," *Moneybox* blog, *Slate*, May 7, 2013, http://www.slate.com/blogs/moneybox/2013/05/07/public_sector_investment_collapse.html (accessed August 12, 2013); and the underlying data at "Real State & Local Consumption Expenditures & Gross Investment, 3 Decimal," *Economic Research—Federal Reserve Bank of St. Louis* (U.S. Department of Commerce: Bureau of Economic Analysis, July 31, 2013), http://research.stlouisfed.org/fred2/series/SLCEC96.
25. "Siemens CEO on US Economic Outlook," *CNBC*, March 14, 2013, http://video.cnbc.com/gallery/?video=3000154454 (accessed August 12, 2013).

26. John Maynard Keynes, *The General Theory of Employment, Interest, and Money*, October 21, 2012, http://ebooks.adelaide.edu.au/k/keynes/john_maynard/k44g/.
27. Peter B. Dixon and Maureen T. Rimmer, "Restriction or Legalization? Measuring the Economic Benefits of Immigration Reform," Cato Institute, August 13, 2009, http://www.cato.org/publications/trade-policy-analysis/restriction-or-legalization-measuring-economic-benefits-immigration-reform (accessed December 14, 2012); Robert Lynch and Patrick Oakford, "The Economic Effects of Granting Legal Status and Citizenship to Undocumented Immigrants," Center for American Progress, March 20, 2013, http://www.americanprogress.org/issues/immigration/report/2013/03/20/57351/the-economic-effects-of-granting-legal-status-and-citizenship-to-undocumented-immigrants/ (accessed August 12, 2013).
28. David Card, "The Impact of the Mariel Boatlift on the Miami Labor Market," Working Paper (National Bureau of Economic Research, August 1989), http://www.nber.org/papers/w3069.
29. Rachel M. Friedberg, "The Impact of Mass Migration on the Israeli Labor Market," *Quarterly Journal of Economics* 116, no. 4 (2001): 1373–1408, doi:10.1162/003355301753265606.
30. Amy Sherman, "Jeb Bush Says Illegal Immigration Is 'Net Zero,'" *Miami Herald*, September 3, 2012, http://www.miamiherald.com/2012/09/01/2980208/jeb-bush-says-illegal-immigration.html.
31. Gordon F. De Jong et al., "The Geography of Immigrant Skills: Educational Profiles of Metropolitan Areas," Brookings Institution, June 9, 2011, http://www.brookings.edu/research/papers/2011/06/immigrants-singer.
32. "State and County QuickFacts," United States Census Bureau, June 27, 2013, http://quickfacts.census.gov/qfd/states/00000.html; Vivek Wadhwa et al., "America's New Immigrant Entrepreneurs: Part I," SSRN Scholarly Paper, Duke Science, Technology & Innovation Paper No. 23 (Rochester, NY: Social Science Research Network, January 4, 2007), http://papers.ssrn.com/abstract=990152.
33. "The 'New American' Fortune 500," Partnership for a New American Economy, June 2011, http://www.renewoureconomy.org/sites/all/themes/pnae/img/new-american-fortune-500-june-2011.pdf.
34. Michael Kremer, "The O-Ring Theory of Economic Development," *Quarterly Journal of Economics* 108, no. 3 (1993): 551–75, doi:10.2307/2118400.
35. Vivek Wadhwa et al., "America's New Immigrant Entrepreneurs: Part I," SSRN Scholarly Paper, Duke Science, Technology & Innovation Paper No. 23 (Rochester, NY: Social Science Research Network, January 4, 2007), http://papers.ssrn.com/abstract=990152; Darrell West, "Inside the Immigration Process," *Huffington Post*, April 15, 2013, http://www.huffingtonpost.com/darrell-west/inside-the-immigration-pr_b_3083940.html (accessed August 12, 2013).
36. Nick Leiber, "Canada Launches a Startup Visa to Lure Entrepreneurs," *Bloomberg Businessweek*, April 11, 2013, http://www.businessweek.com/articles/2013-04-11/canada-launches-a-startup-visa-to-lure-entrepreneurs.
37. Greg Mankiw, "Rogoff Joins the Pigou Club," *Greg Mankiw's Blog*, September

16, 2006, http://gregmankiw.blogspot.com/2006/09/rogoff-joins-pigou-club.html; Ralph Nader and Toby Heaps, "We Need a Global Carbon Tax," *Wall Street Journal*, December 3, 2008, http://online.wsj.com/article/SB122826696217574539.html.
38. P. A. Diamond and E. Saez, "The Case for a Progressive Tax: From Basic Research to Policy Recommendations," *Journal of Economic Perspectives* 25, no. 4 (2011): 165–90.
39. To be more precise, he actually found that on average, higher taxes were correlated with somewhat *faster* growth. See Menzie Chinn, "Data on Tax Rates, by Quintiles," *Econbrowser*, July 12, 2012, http://www.econbrowser.com/archives/2012/07/data_on_tax_rat.html.

Chapter 14 LONG-TERM RECOMMENDATIONS

1. Craig Tomlin, "SXSW 2012 Live Blog Create More Value Than You Capture," *Useful Usability*, March 12, 2012, http://www.usefulusability.com/sxsw-2012-live-blog-create-more-value-than-you-capture/.
2. Sir Winston Churchill and Robert Rhodes James, *Winston S. Churchill: His Complete Speeches, 1897–1963: 1943–1949* (Chelsea House Publishers, 1974), p. 7,566.
3. Martin Luther King, Jr., *Where Do We Go from Here: Chaos or Community?* (New York: Harper & Row, 1967), p. 162.
4. Jyotsna Sreenivasan, *Poverty and the Government in America: A Historical Encyclopedia*, 1st ed. (Santa Barbara, CA: ABC-CLIO, 2009), p. 269.
5. "WGBH American Experience . Nixon | PBS," *American Experience*, http://www.pbs.org/wgbh/americanexperience/features/general-article/nixon-domestic/ (accessed August 12, 2013).
6. Voltaire, *Candide*, trans. Francois-Marie Arouet (Mineola, NY: Dover Publications, 1991), p. 86.
7. Daniel H. Pink, *Drive: The Surprising Truth About What Motivates Us* (New York: Riverhead Books, 2011).
8. Sarah O'Connor, "Amazon Unpacked," *Financial Times*, February 8, 2013, http://www.ft.com/intl/cms/s/2/ed6a985c-70bd-11e2-85d0-00144feab49a.html#slide0.
9. Don Peck, "How a New Jobless Era Will Transform America," *The Atlantic*, March 2010, http://www.theatlantic.com/magazine/archive/2010/03/how-a-new-jobless-era-will-transform-america/307919/?single_page=true.
10. Jim Clifton, *The Coming Jobs War* (New York: Gallup Press, 2011).
11. William Julius Wilson, *When Work Disappears: The World of the New Urban Poor*, 1st ed. (New York: Vintage, 1997).
12. Charles Murray, *Coming Apart: The State of White America, 1960–2010* (New York: Crown Forum, 2013, repr.).
13. Murray argues that harmful changes in values are the most important explanatory factor. As he writes, "The deterioration of social capital in lower-class white America strips the people who live there of one of the main resources through which Americans have pursued happiness. The same may be said of the deterioration in

marriage, industriousness, honesty, and religiosity. These are not aspects of human life that may or may not be important, depending on personal preferences. Together, they make up the stuff of life" (p. 253).

14. Interview with Milton Friedman, *Newsfront*, NET, May 8, 1968; quoted in Gordonskene, "Milton Friedman Explains The Negative Income Tax—1968," *Newstalgia*, December 6, 2011, http://newstalgia.crooksandliars.com/gordonskene/milton-friedman-explains-negative-inco.

15. Raj Chetty et al., "The Economic Impacts of Tax Expenditures: Evidence From Spatial Variation Across the U.S.," White Paper, 2013, http://obs.rc.fas.harvard.edu/chetty/tax_expenditure_soi_whitepaper.pdf.

16. "Citi Community Development Marks National EITC Awareness Day with Release of Money Matters Publication," *News*, Citigroup Inc., January 25, 2013, http://www.citigroup.com/citi/news/2013/130125a.htm.

17. "Gas Guzzler Tax," *Fuel Economy*, United States Environmental Protection Agency, http://www.epa.gov/fueleconomy/guzzler/ (accessed August 12, 2013).

18. "History of the Income Tax in the United States," *Infoplease*, 2007, http://www.infoplease.com/ipa/A0005921.html.

19. Roberton Williams, "The Numbers: What Are the Federal Government's Sources of Revenue?" *The Tax Policy Briefing Book: A Citizens' Guide for the Election, and Beyond* (Tax Policy Center: Urban Institute and Brookings Institution, September 13, 2011), http://www.taxpolicycenter.org/briefing-book/background/numbers/revenue.cfm.

20. In the United States, only income below $113,700 was taxed for Social Security in 2013. See "Social Security and Medicare Tax Rates; Maximum Taxable Earnings," *Social Security: The Official Website of the U.S. Social Security Administration*, February 6, 2013, http://ssa-custhelp.ssa.gov/app/answers/detail/a_id/240/~/social-security-and-medicare-tax-rates%3B-maximum-taxable-earnings.

21. Even when a tax or benefit is nominally paid for by the employer, much of it will ultimately be borne by the employee in the form of lower wages or even lack of employment. See Melanie Berkowitz, "The Health Care Reform Bill Becomes Law: What It Means for Employers," *Monster: Workforce Management*, n.d., http://hiring.monster.com/hr/hr-best-practices/workforce-management/employee-benefits-management/health-care-reform.aspx.

22. Bruce Bartlett, *The Benefit and The Burden: Tax Reform—Why We Need It and What It Will Take* (New York: Simon & Schuster, 2012).

23 Steve Lohr, "Computer Algorithms Rely Increasingly on Human Helpers," *New York Times*, March 10, 2013, http://www.nytimes.com/2013/03/11/technology/computer-algorithms-rely-increasingly-on-human-helpers.html.

24. Jason Pontin, "Artificial Intelligence, With Help From the Humans," *New York Times*, March 25, 2007, http://www.nytimes.com/2007/03/25/business/yourmoney/25Stream.html.

25. Gregory M. Lamb, "When Workers Turn into 'Turkers,' " *Christian Science Monitor*, November 2, 2006, http://www.csmonitor.com/2006/1102/p13s02-wmgn.html.

26. Pontin, "Artificial Intelligence, With Help From the Humans."

27. Daren C. Brabham, "Crowdsourcing as a Model for Problem Solving An Introduction and Cases," *Convergence: The International Journal of Research into New Media Technologies* 14, no. 1 (2008): 75–90, doi:10.1177/1354856507084420.
28. Alyson Shontell, "Founder Q&A: Make a Boatload of Money Doing Your Neighbor's Chores on TaskRabbit," *Business Insider*, October 27, 2011, http://www.businessinsider.com/taskrabbit-interview-2011-10 (accessed August 12, 2013).
29. Tomio Geron, "Airbnb and the Unstoppable Rise of the Share Economy," *Forbes*, January 23, 2013, http://www.forbes.com/sites/tomiogeron/2013/01/23/airbnb-and-the-unstoppable-rise-of-the-share-economy/ (accessed August 12, 2013).
30. Johnny B., "TaskRabbit Names Google Veteran Stacy Brown-Philpot as Chief Operating Officer," *TaskRabbit Blog*, January 14, 2013, https://www.taskrabbit.com/blog/taskrabbit-news/taskrabbit-names-google-veteran-stacy-brown-philpot-as-chief-operating-officer/ (accessed August 12, 2013).
31. Johnny B., "TaskRabbit Welcomes 1,000 New TaskRabbits Each Month," *TaskRabbit Blog*, April 23, 2013, https://www.taskrabbit.com/blog/taskrabbit-news/taskrabbit-welcomes-1000-new-taskrabbits-each-month/.
32. "Employment Situation News Release," Bureau of Labor Statistics, May 3, 2013, http://www.bls.gov/news.release/empsit.htm.

Chapter 15 TECHNOLOGY AND THE FUTURE

1. Charles Perrow, *Normal Accidents: Living with High-Risk Technologies* (Princeton, NJ: Princeton University Press, 1999); *Interim Report on the August 14, 2003 Blackout* (New York Independent System Operator, January 8, 2004), http://www.hks.harvard.edu/hepg/Papers/NYISO.blackout.report.8.Jan.04.pdf.
2. Steven Cherry, "How Stuxnet Is Rewriting the Cyberterrorism Playbook," *IEEE Spectrum* podcast, October 13, 2010, http://spectrum.ieee.org/podcast/telecom/security/how-stuxnet-is-rewriting-the-cyberterrorism-playbook.
3. Bill Joy, "Why the Future Doesn't Need Us," *Wired*, April 2000, http://www.wired.com/wired/archive/8.04/joy_pr.html.
4. The costs of gene sequencing are dropping even more quickly than those of computing. A comprehensive discussion of the genomics revolution is far beyond the scope of this book; we mention it here simply to highlight that it is real, and likely to bring profound changes in the years and decades to come. See Kris Wetterstrand, "DNA Sequencing Costs: Data from the NHGRI Genome Sequencing Program (GSP)," National Human Genome Research Institute, July 16, 2013, http://www.genome.gov/sequencingcosts/.
5. On gaming, see Nicholas Carr, *The Shallows: What the Internet Is Doing to Our Brains* (New York: W. W. Norton & Company, 2011); on cyberbalkanization, see Marshall van Alstyne and Erik Brynjolfsson, "Electronic Communities: Global Villages or Cyberbalkanization?" *ICIS 1996 Proceedings*, December 31, 1996, http://aisel.aisnet.org/icis1996/5; and Eli Pariser, *The Filter Bubble: How the New Personalized Web Is Changing*

What We Read and How We Think (New York: Penguin, 2012); on social isolation see Sherry Turkle, *Alone Together: Why We Expect More from Technology and Less from Each Other* (New York: Basic Books, 2012); and Robert D. Putnam, *Bowling Alone: The Collapse and Revival of American Community*, 1st ed. (New York: Simon & Schuster, 2001); finally, on environmental degradation, see Albert Gore, *The Future: Six Drivers of Global Change, 2013.*

6. Chad Brooks, "What Is the Singularity?" *TechNewsDaily*, April 29, 2013, http://www.technewsdaily.com/17898-technological-singularity-definition.html.

7. To improve the odds that he will be alive to see the singularity (he'll be ninety-seven in 2045), Kurzweil has put himself on a self-engineered diet that includes taking 150 nutritional supplements every day. See Kristen Philipkoski, "Ray Kurzweil's Plan: Never Die," *Wired*, November 18, 2002, http://www.wired.com/culture/lifestyle/news/2002/11/56448.

8. Steve Lohr, "Creating Artificial Intelligence Based on the Real Thing," *New York Times*, December 5, 2011, http://www.nytimes.com/2011/12/06/science/creating-artificial-intelligence-based-on-the-real-thing.html.

9. Gareth Cook, "Watson, the Computer *Jeopardy!* Champion, and the Future of Artificial Intelligence," *Scientific American*, March 1, 2011, http://www.scientificamerican.com/article.cfm?id=watson-the-computer-jeopa.

10. Martin Luther King Jr., "Sermon at Temple Israel of Hollywood," February 26, 1965, http://www.americanrhetoric.com/speeches/mlktempleisraelhollywood.htm.

ILLUSTRATION SOURCES

Figure 1.1 and 1.2 Human Social Development Index figures from Ian Morris, *Why the West Rules . . . For Now: The Patterns of History, and What They Reveal About the Future* (New York: Picador, 2011).
Worldwide human population figures are an average of estimates from the U.S. Census Bureau's "Historical Estimates of World Population," http://www.census.gov/population/international/data/worldpop/table_history.php.
World population for 2000 from the CIA World Factbook

3.1 Author's own

3.2 Author's own

3.3 Supercomputer speeds:
http://www.riken.jp/en/pr/publications/riken_research/2006/
http://www.intel.com/pressroom/kits/quickrefyr.htm
http://www.green500.org/home.php
Hard drive cost:
http://www.riken.jp/en/pr/publications/riken_research/2006/
http://www.intel.com/pressroom/kits/quickrefyr.htm
http://www.green500.org/home.php
Supercomputer energy efficiency:
http://ed-thelen.org/comp-hist/CRAY-1-HardRefMan/CRAY-1-HRM.html
http://www.green500.org/home.php
Transistors per chip:
http://www.intel.com/pressroom/kits/quickrefyr.htm
Download speed:
http://www.akamai.com/stateoftheinternet/

7.1 U.S. Bureau of Economic Analysis

7.2 Chad Syverson, "Will History Repeat Itself? Comments on 'Is the Information Technology Revolution Over?'," *International Productivity Monitor* 25 (2013), 37–40. John W. Kendrick, "Productivity Trends in the United States," National Bureau of Economic Research, 1961. David M. Byrne, Stephen D. Oliner, and Daniel E. Sichel, "Is the Information Technology Revolution Over?," *International Productivity Monitor* 25 (Spring 2013), 20–36.

9.1 http://research.stlouisfed.org/fred2/graph/?id=USARGDPC
http://www.census.gov/hhes/www/income/data/historical/people/

9.2 D. Acemoglu and David Autor, "Skills, tasks and technologies: Implications for employment and earnings," *Handbook of Labor Economics* 4 (2011), 1043–1171.
9.3 http://research.stlouisfed.org/fred2/graph/?id=GDPCA
http://research.stlouisfed.org/fred2/graph/?id=A055RC0A144NBEA
http://research.stlouisfed.org/fred2/graph/?id=W270RE1A156NBEA

10.1 N/A

11.1 http://research.stlouisfed.org/fred2/series/USPRIV
http://research.stlouisfed.org/fred2/graph/?id=USARGDPH

INDEX

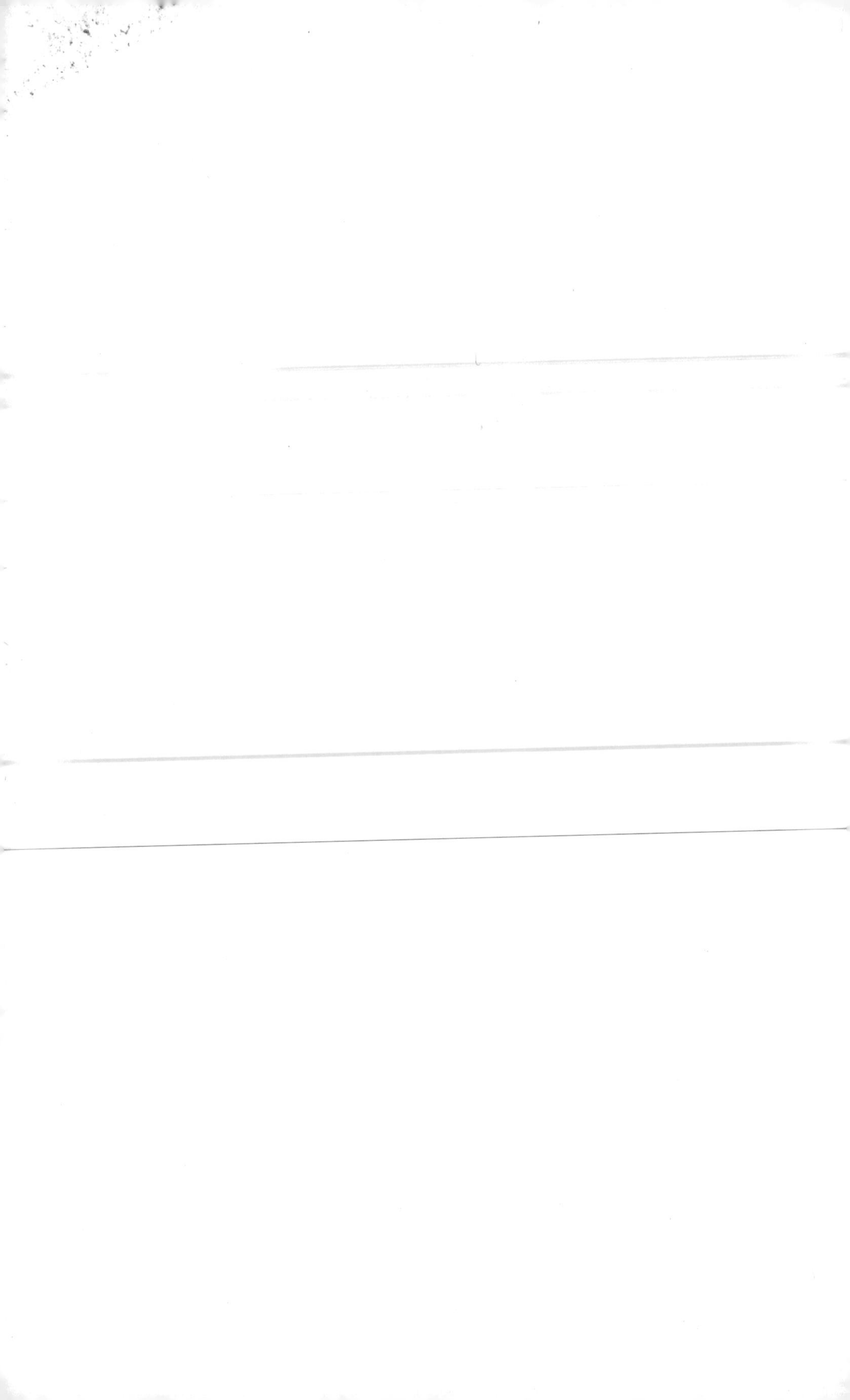